Carbon Nanostructures

More information about this series at http://www.springer.com/series/8633

K. Sridharan · B. Srinivasu · Vikramkumar Pudi

Low-Complexity Arithmetic Circuit Design in Carbon Nanotube Field Effect Transistor Technology

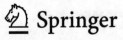 Springer

K. Sridharan
Department of Electrical Engineering
Indian Institute of Technology Madras
Chennai, India

Vikramkumar Pudi
Department of Electrical Engineering
Indian Institute of Technology Tirupati
Tirupati, Andhra Pradesh, India

B. Srinivasu
School of Computing and Electrical
Engineering
Indian Institute of Technology Mandi
Mandi, Himachal Pradesh, India

ISSN 2191-3005   ISSN 2191-3013   (electronic)
Carbon Nanostructures
ISBN 978-3-030-50701-5   ISBN 978-3-030-50699-5   (eBook)
https://doi.org/10.1007/978-3-030-50699-5

This Springer imprint is published by the registered company Springer Nature Switzerland AG
The registered company address is: Gewerbestrasse 11, 6330 Cham, Switzerland

Dedicated to our families

Preface

Overview

This book grew out of our research enquiry into arithmetic in emerging device technologies. In particular, it presents our studies beginning in summer of 2012 at the Indian Institute of Technology Madras, on design of low-complexity arithmetic circuits in Carbon Nanotube Field Effect Transistor (CNTFET) technology. The focus has been on obtaining designs with low transistor count. Taking advantage of the large number of unary operators of multi-valued logic, the book explores careful selection of the operators to obtain efficient designs. Simulations of the circuits validate the proposed designs.

Organization and Features

Chapter 1 presents the motivation for the work described in this book. This chapter also gives an overview of the literature on the subject. Chapter 2 presents terminology pertaining to carbon nanotube field effect transistors and ternary arithmetic. Special attention is given to unary operators of multivalued logic. Chapter 3 introduces the reader to CNTFET-based design of basic logic elements. Chapter 4 presents direct and efficient single ternary digit adder designs. Chapter 5 extends the material in Chap. 4 by addressing multi-ternary digit adder design in CNTFET technology. Chapter 6 proceeds to explore CNTFET-based multiplier design. Chapter 7 presents simulation results for the various designs. Chapter 8 discusses automatic synthesis of CNTFET-based logic circuits. Chapter 9 presents a summary of the work described in the book and outlines extensions. HSPICE programs for simulation are presented in an appendix. Python code for automatic synthesis is also presented in a second appendix.

Audience

This book presents material that is appropriate for courses at the senior under-graduate level and graduate level in the areas of nanoelectronics, computer arithmetic and embedded systems. It can also be used as a supplement to courses on digital circuits and laboratories on digital systems. The book is also suitable for researchers in the areas of computer arithmetic, nanotechnologies and VLSI design. In addition, the book provides examples in HSPICE for simulation of the designs. Besides, the reader is introduced to automatic synthesis via a Python snippet. Basic familiarity with logic design is adequate to follow the material presented in this book.

Chennai, India   K. Sridharan
Mandi, India   B. Srinivasu
Tirupati, India   Vikramkumar Pudi

Acknowledgments The authors owe a word of thanks to many people who helped in various ways. The authors thank their families and friends for their support. Special thanks go to Dr. Mayra Castro and Dr. Thomas Ditzinger, Springer editors, for obtaining reviews for chapters in this book. Thanks also to the anonymous reviewers for the comments. The authors would also like to thank Dr. Guido Zosimo-Landolfo and Dr. Dieter Merkle of Springer for their assistance. Thanks also to Ms. Jialin Yan, Mr. Dinesh Natarajan, Ms. Eva Schoeler and Ms. Priyadharshini Subramani of Springer for administrative assistance. The authors also acknowledge the support of Indian Institute of Technology Madras, Indian Institute of Technology Mandi, Indian Institute of Technology Tirupati and Nanyang Technological University, Singapore.

About This Book

Carbon is known to be a powerful alternative to silicon for various applications. Research in electronics involving carbon nanotubes has been reported since the 1960s. A carbon nanotube exhibits interesting behaviour. It can act either as a metal or as a semi-conductor depending on a parameter called the chirality. When the chirality is set to allow the tube to function as a semi-conductor, it can be used as the channel material for a transistor.

This book introduces the reader to Carbon Nanotube Field Effect Transistor (CNTFET) technology, an emerging nanotechnology and an alternative to CMOS. It then examines the problem of designing efficient arithmetic circuits in CNTFET technology. CNTFETs provide the possibility of realizing two distinct threshold voltages merely by use of different diameters of the carbon nanotube.

Starting with design of basic logic circuits, the book proceeds to discuss efficient design of CNTFET-based single ternary digit adders. The design of multi-ternary digit adders in CNTFET technology is then studied. This is followed by the design of a ternary multiplier in CNTFET. Detailed simulation results are also presented for all the circuits. We also discuss automation of the synthesis process.

Contents

About the Authors

K. Sridharan received his Ph.D. from Rensselaer Polytechnic Institute, Troy, New York in 1995. He was an Assistant Professor at Indian Institute of Technology (IIT) Guwahati from 1996 to 2001. Since June 2001, he is with IIT Madras where he is presently a Professor. He was a visiting faculty member at Nanyang Technological University (NTU), Singapore in 2000–2001 and 2006–2008. He has supervised 6 Ph.Ds. and holds 2 patents. He is an author of two books published by Springer in the areas of robotics and quantum dot cellular automata. He has also authored/co-authored approximately 100 papers in various journals and conferences. He is a recipient of the 2009 Dr. Vikram Sarabhai Research Award for his contributions to electronics, telematics, informatics and automation. He also received the Tan Chin Tuan fellowship in Engineering for research in nanoelectronics and VLSI at Nanyang Technological University in 2011. He was an Associate Editor of the IEEE Transactions on Industrial Electronics from 2012 until April 2019. He is a Fellow of IETE (India), IE (India) and a Senior Member of IEEE. His research interests include robotics, embedded systems and digital design in emerging nanotechnologies.

B. Srinivasu received his Ph.D. from Indian Institute of Technology Madras in 2017. He was a postdoctoral research fellow at the School of Computer Science and Engineering, Nanyang Technological University between May 2017 and July 2019. He is currently an Assistant Professor at the School of Computing and Electrical Engineering, Indian Institute of Technology Mandi. His research interests include carbon nanotube field effect transistor technology and digital fingerprinting.

Vikramkumar Pudi received his Ph.D. from Indian Institute of Technology Madras in 2014. He was a postdoctoral fellow with the Microsystems and Nanotechnology group at the University of British Columbia, Canada in 2014–2015. During the period from 2015 to 2018, he was with Nanyang Technological

University, Singapore as a Research Fellow. He is currently an Assistant Professor at Indian Institute of Technology Tirupati. He is the author of a Springer book on quantum dot cellular automata. His research interests include digital design, cyber security and cryptography and nanoelectronics.

Chapter 1
Introduction

There has been tremendous growth in the semiconductor industry in the last few decades. This has largely followed Gordon Moore's prediction in the 1960s that number of transistors per chip would double every two years approximately. However, there have been concerns about scaling limits of silicon MOSFETs during the last twenty years [1]. As a consequence, variations of the basic theme have been suggested and two categories have emerged, one of which attempts further miniaturization while the other aims at diversification. It is becoming evident from a recent International Technology Roadmap for Semiconductors' report [2] that further transistor scaling may not be viable. There have been attempts to manufacture chips by shifting the device structure from horizontal to vertical and building multiple layers of circuitry, one on top of another.

Recent developments in silicon MOSFET include the Double Gate Field Effect Transistor (DGFET) [3] and the Fin Field Effect Transistor (FinFET) [4]. There have also been efforts on using new device materials for making transistors. One approach in this direction is based on carbon. Unique electronic, thermal and mechanical properties of carbon have enabled its use in diverse applications. The early work of Mildred Dresselhaus [5] has paved the way for electronics made from tubes, wires and sheets of carbon. It is well-known that carbon permits *hybridization* styles such as sp, sp^2 and sp^3 [6]. The sp^2 hybridization in carbon leads to *graphite* which can function as a conductor. This book concerns *graphene* which corresponds to one atomic layer of graphite. In particular, single-walled carbon nanotubes constructed by rolling up a single graphene sheet are valuable for fabricating transistors.

A carbon nanotube exhibits interesting behaviour. It can act either as a metal or as a semi-conductor depending on a parameter termed as the *chirality*. When the chirality is set to allow the tube to function as a semi-conductor, it can be used as the channel material for a transistor. A Field Effect Transistor (FET) that has one or more carbon nanotubes as the channel constitutes what is called as a Carbon Nanotube Field Effect Transistor (CNTFET for short).

© Springer Nature Switzerland AG 2020
K. Sridharan et al., *Low-Complexity Arithmetic Circuit Design in Carbon Nanotube Field Effect Transistor Technology*, Carbon Nanostructures,
https://doi.org/10.1007/978-3-030-50699-5_1

1.1 Motivation

The broad objective of this work is to explore the possibility of making microprocessors using CNTFETs. To this end, the book attempts to handle one aspect of a traditional central processing unit (CPU), namely the arithmetic unit. In particular, the book concentrates on arithmetic circuit design in CNTFET technology. The goal is to obtain low complexity designs. We briefly trace the history in the direction of digital design to provide the context for the work presented in this book.

1.1.1 Digital Design and Fabrication in CNTFET Technology

Digital design in CNTFET as such is not new. There are some efforts on binary logic-based circuits in CNTFET [7]. The authors in [7] provide details of logic gates fabricated using CNTFETs. Probabilistic analysis and design of digital circuits particularly taking into account some issues at the fabrication level have been studied in [8]. A single instruction computer based on the classical *SUBNEG* instruction (that performs a subtract and branch if negative operation) has been described in [9]. All the CNTFETs and interconnections are predetermined during the design. Recently, a 16-bit microprocessor based on complementary carbon nanotube transistors has been reported [10].

1.1.2 Beyond Binary

One of the early efforts suggesting the suitability of ternary logic for CNTFET-based designs is [11]. The work presented in this book draws upon the characteristics of carbon nanotubes for multivalued logic circuit design [11]. In particular, carbon nanotubes have bandgaps that depend on the diameter of the tubes. Further, the bandgap turns out to be a measure of the threshold voltage of the CNTFET. In particular, CNTFETs provide the possibility of realizing two distinct threshold voltages merely by use of different diameters of the carbon nanotube. Taking advantage of this, ternary logic-based design of single digit adders has been actively pursued recently [12, 13]. However, there has been only limited work beyond one-digit adder designs.

1.2 Challenges in CNTFET-Based Digital Design

Direct design of ternary logic circuits typically leads to a large number of transistors. One of the challenges in digital design in CNTFET technology is coming up with a design that has a fairly small transistor count.

Binary as well as ternary logic are characterised by unary operators [14]. While there are only four unary operators in binary, there are as many as twenty seven unary operators in ternary. It is therefore a challenge to choose the appropriate set of operators to obtain an efficient design.

Identifying an adder/multiplier configuration that will perform well in ternary logic (when realized in CNTFET technology) is also an interesting problem. In general, adders (or multipliers) that perform well in one technology may not necessarily do well in another.

1.3 Objectives of the Book

The objectives of the book are as follows.

1. Explore the various unary operators in ternary logic and design low-complexity circuits using CNTFETs.
2. Design efficient CNTFET-based single ternary-digit *(trit)* adders *by careful selection of unary operators* and compare with existing designs with respect to device count.
3. Design a multi-trit adder with low design complexity in CNTFET technology.
4. Extend the unary operator-based approach to the design of a low-complexity single-trit multiplier in CNTFET.
5. Present extensive simulation studies on CNTFET-based circuits.
6. Consider the task of automating the synthesis process.

1.4 Contributions of the Book

The book presents ternary arithmetic circuits in CNTFET technology. In particular, the contributions are described in the sections to follow.

1.4.1 Efficient Design of CNTFET-Based Ternary Half-Adder and Full-Adder

The book first examines efficient designs for one (ternary) digit addition in CNTFET. In particular, designs that lead to low device-count are developed by careful selection of unary operators. Prior work on single digit addition has been largely either decoder-based or using specific unary operators [11, 15] without emphasis on reduction of device count. We present modifications that enable efficient realization of single digit adders.

1.4.2 Efficient Design of a CNTFET-Based Ternary Multi-digit Adder

To our knowledge, prior work on ternary multi-digit adders in CNTFET is limited. We develop a ternary multi-trit adder based on the notion of conditional sum. In the process, we present two modified half-adder designs (using unary operators).

1.4.3 Efficient Design of a Ternary Multiplier in CNTFET Technology

The book also presents the design of a ternary multiplier in CNTFET technology. It is observed that, in the ternary setting, multiplication of a pair of single-trit numbers leads to two components in general (namely a product and a carry). We discuss a direct approach and then present a low-complexity design for multiplication.

1.4.4 Extensive Simulation Studies

While the primary focus of the book is on design of addition and multiplication circuits in CNTFET, it is also of interest to simulate the designs.

1.4.5 Automatic Sythesis of Ternary Logic Circuits

One question that is natural to ask is the following: *Can we automate the synthesis process ?* We answer this in the affirmative in this book and outline an algorithm for synthesis that combines a geometrical representation with unary operators of multi-valued logic. The geometrical representation is motivated by early work described in [16].

1.5 Literature Survey

In this section, we discuss prior work on multivalued logic as well as digital design in CNTFET technology.

1.5.1 Multiple Valued Logic

Research on multiple valued logic has been steadily growing in the last few decades. The advantages with respect to interconnect complexity increase as we move towards multivalued logic. An economical radix is ternary corresponding to a base of 3. A ternary arithmetic unit is presented in [17]. The advantages of multiple valued logic with respect to reduction of active devices in the hardware and consequently reduction in the power consumption are reported in [18]. The authors in [18] present a 32 × 32 signed-digit multiplier in multiple valued logic and it is reported that there is a 50% savings in the power dissipation over a binary multiplier of the same size. The design of basic ternary gates as well as circuits such as JK flip flop and T-flip flop using MOS transistors are reported in [19, 20]. CMOS-based ternary circuits are presented in [21] and the designs are based on PMOS and NMOS transistors and resistors. The authors also suggest designs with pass transistor logic replacing the physical resistor with a MOSFET. A ternary full adder adder in CMOS is presented in [22]. A 5×5 ternary multiplier, designed in CMOS, is reported in [23] where the authors use quasi-adiabatic ternary logic. However, as the continuous scaling of CMOS continues, the sub-threshold effects in the CMOS (such as OFF current and leakage power) increase. Hence, at the nano-scale, it is not clear how well CMOS can handle multiple-valued logic.

1.5.2 Design of Ternary Logic Circuits in CNTFET Technology

Given the possibility of realizing multiple threshold voltages in a CNTFET, it is of interest to discuss prior work on design of ternary circuits. A ternary logic-based approach for various circuits in CNTFET is proposed in [11], where the authors use resistive logic to implement a few unary operators. The authors also present the design of a ternary half adder and a single-digit multiplier. This design uses large on-chip resistors which consume substantial area as well as high power. An alternate approach to implement ternary circuits has been adopted by some researchers. It is based on converting the ternary signals into binary, realizing a binary circuit and finally, encoding the outputs in ternary. One example of work in this direction is [12] where the authors have presented arithmetic circuits in ternary via a ternary decoder/encoder pair (for the conversion). This paper also presents a ternary half-adder and single digit multiplier using NAND and NOR gates. The authors in [24] present an alternative to the ternary NAND and NOR which requires lesser number of CNTFETs. The authors in [15] use a standard ternary inverter reported in [25] to differentiate logic 1 from logic 0 and logic 2. A ternary full adder based on this idea is presented which takes less hardware compared to the one with the decoder.

Not much is known on ternary logic circuits beyond designs for specific operations. In particular, generic synthesis strategies appear to be scarce although ternary function minimization has been briefly explored as early as 1968 [16]. In particular, a graphical approach for three variables based on a ternary cube is reported in [16] where the author also presents algebraic relationships to facilitate minimization of functions of four (or more) variables. However, the focus in [16] is only on function minimization and not on the circuit realization. Besides arithmetic circuits, there has been some research dating back to the 1970s on ternary memory units [19].

There are some recent efforts in the direction of low-power design. These include a low-power ternary full-adder in [26] and low-power ternary multipliers in [27, 28] and [29]. Designs with low power-delay product are also being currently studied [30, 31].

1.6   Organization of the Book

Next chapter presents CNTFET terminology appropriate for the remainder of the book.

In Chap. 3, we present efficient designs of basic ternary logic circuits.

Chapter 4 presents CNTFET-based designs of single digit adders including a half adder and a full adder using the unary operators.

In Chap. 5, we develop an efficient CNTFET-based multi-trit adder design. It is based on efficient single-digit ternary adder designs.

Chapter 6 is devoted to low-complexity multiplier design in CNTFET technology.

Chapter 7 presents detailed simulation studies for the designs discussed in the earlier chapters.

Automatic synthesis of CNTFET-based circuits is outlined in Chap. 8.

Chapter 9 summarizes the contributions of the book and indicates the scope for future work.

Sample HSPICE programs for simulation of the designs are presented in the first appendix. A Python program for automatic synthesis is presented in the second appendix.

1.7   Summary

In this chapter, we have presented a brief introduction to an emerging nanotechnology based on carbon nanotubes. We have discussed the advantages of ternary logic and some design issues. In the next chapter, we introduce CNTFET terminology appropriate for the remainder of the book.

References

1. Frank, D.J., Dennard, R.H., Nowak, E., Solomon, P.M., Tuar, Y., Wong, H.S.P.: Device scaling limits of Si MOSFETs and their application dependencies. Proc. IEEE **89**(3), 259–288 (2001)
2. Gargini, P.: ITRS—Past, Present and Future (2015). http://www.itrs2.net/itrs-reports.html
3. Balestra, F., Cristoloveanu, S., Benachir, M., Elwa, T.: Double-gate silicon-on-insulator transistor with volume inversion : a new device with greatly enhanced performance. IEEE Electron Devices Lett. **8**(9), 410–412 (1987)
4. Huang, X., Lee, W.-C., Ku, C., Hisamoto, D., Chang, L., Kedzierski, J., Anderson, E., Takeuchi, H., Choi, Y.-K., Asano, K., Subramanian, V., King, T.J., Bokor, J., Hu, C.: Sub 50-nm FinFET: PMOS. IEDM Tech. Digest, 67–70 (1999)
5. Dresselhaus, M., Mavroides, J.G.: The Fermi structure of graphite. IBM J. Res. Develop. **8**(3), 262–267 (1964)
6. Atkins, P., de Paula, J.: Physical Chemistry, 9th edn. W.H. Freeman (2009)
7. Bachtold, A., Hadley, P., Nakanishi, T., Dekker, C.: Logic circuits with carbon-nanotube transistors. Science **294**(5545), 1317–1320 (2001)
8. Zhang, J., Patil, N., Mitra, S.: Probabilistic Analysis and design of metallic-carbon-nanotube-tolerant digital logic circuits. IEEE Trans. Comput.-Aided Des. Integr. Circuits Syst. **28**(9), 1307–1320 (2009)
9. Shulaker, M.M., Hills, G., Patil, N., Wei, H., Chen, H.-Y., Wong, H.-S.P., Mitra, S.: Carbon nanotube computer. Nature **501**, 526–535 (2013)
10. Hills, G., Lau, C., Wright, A., Fuller, S., Bishop, M.D., Srimani, T., Kanhaiya, P., Ho, R., Amer, A., Stein, Y., Murphy, D., Arvind, Chandrakasan, A., Shulaker, M.M.: Modern microprocessor built from complementary carbon nanotube transistors. Nature **572**, 595–602 (2019)
11. Raychowdhury, A., Roy, K.: Carbon-Nanotube-based voltage-mode multiple-valued logic design. IEEE Trans. Nanotechnol. **4**(2), 168–179 (2005)
12. Lin, S., Kim, Y.B., Lombardi, F.: CNTFET-based design of ternary logic gates and arithmetic circuits. IEEE Trans. Nanotechnol. **10**(2), 217–225 (2011)
13. Moaiyeri, M.H., Doostaregan, A., Navi, K.: Design of energy-efficient and robust ternary circuits for nanotechnology. IET Circuits, Devices, Syst. **5**(4), 285–296 (2011)
14. Miller, D.M., Thornton, M.A.: Multiple Valued Logic: Concepts and Representations. Morgan and Claypool Publishers (2008)
15. Keshavarzian, P., Sarikhani, R.: A novel CNTFET-based ternary full adder. Circuits, Syst. Signal Process. **33**, 665–679 (2014)
16. Hurst, S.L.: An extension of binary minimization techniques to ternary equations. Comput. J. **11**(3), 277–286 (1968)
17. Halpern, I., Yoeli, M.: Ternary arithmetic unit. Proc. IEE **115**(10), 1385–1388 (1968)
18. Kameyama, M., Kawahito, S., Higuchi, T.: A Multiplier Chip with multiple-valued bidirectional current-mode logic circuits. IEEE Comput. **23**(4), 43–56 (1988)
19. Mouftah, H.T., Jordan, I.B.: Design of ternary COS/MOS memory and sequential circuits. IEEE Trans. Comput. 281–288 (1977)
20. Mouftah, H.T., Smith, K.C.: Design and implementation of three-valued logic systems with MOS integrated circuits. IEE Proc. Part G **127**(4), 165–168 (1980)
21. Wu, X.W., Prosser, F.P.: CMOS ternary logic circuits. IEEE Proc. Circuits, Devices Syst. **137**(1), 21–27 (1990)
22. Srivastava, A., Venkatapathy, K.: Design and implementation of a low power ternary full adder. VLSI Design **4**(1), 75–81 (1996)
23. Mateo, D., Rubio, A.: Design and implementation of a 5 × 5 trits multiplier in a quasi-adiabatic ternary CMOS logic. IEEE J. Solid-State Circuits **33**(7), 1111–1116 (1998)
24. Navi, K., Rashtian, M., Khatir, A., Keshavarzian, P.: High speed capacitor-inverter based carbon nanotube full adder. Nanoscale Res. Lett. **5**, 859–862 (2010)
25. Lin, A., Patil, N., Wei, H., Mitra, S., Wong, H.S.P.: ACCNT—a Metallic-CNT-tolerant design methodology for carbon-nanotube VLSI : concepts and experimental demonstration. IEEE Trans. Electron Devices **56**(12), 2969–2978 (2009)

26. Sharma, T., Kumre, L.: CNTFET based design of ternary arithmetic modules. Circuits, Syst. Signal Process. **38**, 4640–4666 (2019)
27. Sahoo, S.K., Dhoot, K., Sahoo, R.: High performance ternary multiplier using CNTFET. In: Proceedings of 2018 IEEE Computer Society Annual Symposium on VLSI, pp. 269–274 (2018)
28. E. Shahrom and S.A. Hosseini. A new low power multiplexer based ternary multiplier using CNTFETs. Int. J. Electron. Commun. (AEÜ)
29. Sharma, T., Kumre, L.: Design of low power multi-ternary digit multiplier in CNTFET technology. Microprocess. Microsyst. **73**, 1–8 (2020)
30. Jaber, R.A., Kassem, A., El-Hajj, A.M., El-Nimri, L.A., Haidar, A.M.: High-performance and energy-efficient CNFET-based designs for ternary logic circuits. IEEE Access **7**, 93871–93886 (2019)
31. Zarandi, A.D., Reshadinezhad, M.R., Rubio, A.: A systematic method to design efficient ternary high performance CNTFET-based logic cells. IEEE Access **8**, 58585–58593 (2020)

Chapter 2
Basics of CNTFET and Ternary Logic

In this chapter, we present key aspects of the CNTFET technology. As indicated earlier, carbon nanotubes have bandgaps that are dependent on the diameter of the tubes [1]. Also, the bandgap turns out to be a measure of the threshold voltage of the CNTFET. Variation in diameter of a carbon nanotube can be related to different threshold voltages thereby permitting use of multi-valued logic for designs in CNT-FET technology. We therefore begin by discussing about carbon nanotubes and then proceed to terminology related to transistors made with carbon nanotubes.

2.1 Carbon Nanotube (CNT)

Carbon Nanotubes (CNTs) are tubular structures of carbon atoms in sp^2 hybridization (the sp^2 hybridization involves mixing of $1s$ and $2p$ atomic orbitals). The diameter of these tubes is in the order of nanometers while the tubes themselves can be a few micrometers long. CNTs are formed by rolling a one-dimensional graphene sheet. The axis of rolling is commonly known as the *"chiral vector"* indicated by $\vec{C}$. It is expressed by Eq. (2.1).

$$\vec{C} = ma_1 + na_2 \tag{2.1}$$

Here, (m,n) is known as the *'chirality'* of the CNT while a_1 and a_2 are the unit vectors of the two-dimensional graphene sheet as indicated in Fig. 2.1. A graphene sheet can be rolled in various ways. Depending upon the axis of rolling, they are referred to as *chiral, armchair* or *zig-zag CNT* as shown in Fig. 2.1. The vector (m, n) indicates the way of wrapping a graphene sheet to follow a hollow cylindrical tube. The wrapping determines the behaviour of the tube: in particular, the formula $m - n = 3i + p$ (where i and p are integers) holds [2] and the classification is as follows

© Springer Nature Switzerland AG 2020
K. Sridharan et al., *Low-Complexity Arithmetic Circuit Design in Carbon Nanotube Field Effect Transistor Technology*, Carbon Nanostructures,
https://doi.org/10.1007/978-3-030-50699-5_2

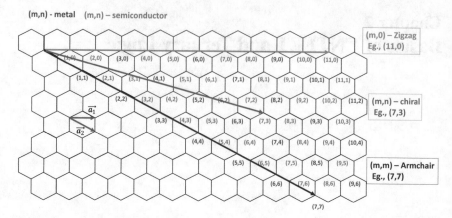

Fig. 2.1 Graphene lattice structure

- p = 0: metallic CNT
- p = 1, 2: semiconducting CNT
- m−n = 0: metallic CNT

We can also relate the metal/semi-conducting behavior to the different CNT types as follows.

- Armchair CNT: (10, 10): metallic CNT ($m = n$)
- Zig-zag CNT: (18, 0): metallic CNT ($p = 0$)
- Chiral CNT: (19, 0): semiconducting CNT ($p = 1$)

We use semiconducting CNTs in our work. In particular, chirality of (19,0) is of interest (as we will see later). The diameter of the CNT can be derived in terms of the chirality of the CNT as given in Equation (2.2).

$$D_{CNT} = \frac{\sqrt{3}a_0}{\pi}\sqrt{m^2 + mn + n^2} \qquad (2.2)$$

Here a_0 is the inter atomic distance between each carbon atom, (which is 0.142 nm or 1.42 A^0) and m, n are as defined earlier. A CNT with a chiral vector of (19, 0) has a diameter of 1.487 nm. The CNT can be grown via different methods. These include (i) arc-discharge method (ii) laser-ablation technique and (iii) chemical vapor deposition (CVD) method. Each of these methods can generate both single-walled nanotubes (SWNT) and multi-walled nanotubes (MWNT). One of the early efforts in obtaining an MWNT using the arc-discharge method was by Iijima [3]. In the CNT growth process, typically one-third are metallic CNTs while the remaining (two-third) are semi-conducting CNTs. The advancements in the CNT growth process have lead to an increase in the percentage of semiconducting CNTs produced. For applications involving CNTs in transistors, the metallic CNTs need to be removed and this is accomplised by a dedicated removal process. We discuss more about transistors using CNTs in the next section.

2.2 Carbon Nanotube Field Effect Transistor (CNTFET)

Field-Effect transistors have been fabricated with one semiconducting single-wall carbon nanotube (SWNT) connected to two metal electrodes as early as 1998 [5]. The channel in the conventional MOSFET is replaced with a semiconducting CNT in a CNTFET while the source and drain can be a metal or a metallic CNT. Depending on the type of source/drain material, it is categorized as (i) MOSFET-like CNTFET (MOSFET-CNTFET) [6]. (ii) Schottky Barrier CNTFET (SB-CNTFET) and (iii) Band-to-Band tunneling CNTFET (BTBT-CNTFET).

1. Schottky Barrier CNTFET (SB-CNFET): The structure of the SB-CNTFET is shown in Fig. 2.2, where the two metallic contacts to the SWNT act as source and drain. The operation of this not only depends on the channel potential but also depends on changing the source and drain materials [7]. The difference of the energy gap between the metallic contact and SWNT leads to a Schottky barrier at the contacts. So different heights of the Schottky barrier are possible for several metal types with various diameters. The typically used materials for the metallic contacts are Al, Pd and Ti. A complete compact model of the SB-CNTFET is presented in [8]. The presence of the Schottky barrier leads to ambipolar characteristics.
2. MOSFET-like CNTFET: The device structure is shown in Fig. 2.3, where doped CNTs are used as source and drain. The structure is similar to the conventional MOSFET, thus the name MOSFET-like CNTFET. The problems of having ambipolar behavior in the SB-CNTFET can be overcome in the MOSFET-CNTFET and hence the higher ON current due to the absence of the Schottky barrier at the drain-to-SWNT. Since the device is similar to the conventional Si-MOSFET, the logic circuits with the MOSFET can be replaced with a CNTFET. A compact model of the MOSFET-like CNTFET is reported in [9, 10].

Fig. 2.2 Schottky Barrier type CNTFET [4]

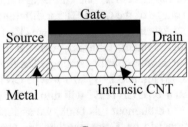

Fig. 2.3 MOSFET-like CNTFET [4]

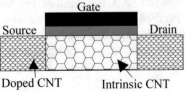

3. Band to Band tunneling or Tunneling CNTFET (BTBT-CNTFET or T-CNTFET): In an SB-CNTFET, the charge transfer is via injection of carriers through a metal-SWNT interface while in MOSFET-like CNTFET, the operation is through the modulation of the charge density. In a T-CNTFET, the tunneling of the carriers takes place at the source and drain contacts. The heavily doped contacts such as p^+ and n^+ act as source and drain while an intrinsic CNT acts as a channel. This type of device is more suitable for low-energy applications. The disadvantage of the T-CNTFET is the limitation on the ON current, which depends on the tunneling of the carriers which is below the range required to build a logic circuit.

In this book, the logic circuits are designed using MOSFET-like CNTFET. We will refer to a MOSFET-like CNTFET as merely CNTFET henceforth. In a CNTFET, the threshold voltage is dependent on the diameter of the CNT as indicated earlier. The threshold voltage of a CNTFET is given by

$$V_{th} = \frac{\sqrt{3}}{3}\frac{aV_\pi}{eD_{CNT}} \tag{2.3}$$

where a is the lattice constant (carbon to carbon atom distance) whose (numerical) value is 2.49 A^0. V_π is carbon $\pi - \pi$ bond energy in the tight bonding model (and this is equal to 3.033eV) while e is the unit electron charge and D_{CNT} is the CNT diameter given by (2.2). Hence, we have

$$V_{th} = \frac{0.42}{D_{CNT}(nm)}V \tag{2.4}$$

The threshold voltage of the CNTFET can be calculated from the diameter of the CNT. In particular, the threshold voltage is inversely proportional to the diameter of the CNT. Hence, by changing the chirality of the CNT, different threshold voltages can be obtained. In effect, taking advantage of chirality-dependent threshold voltage of the CNTFET, one can design multi-valued logic digital circuits. The threshold voltage of the CNTFET for different chiralities is reported in Table 2.1.

Assuming a supply voltage (V_{dd}) of 0.9 V, we have logic 1 as 0.45 V in a ternary logic system. Table 2.1 gives the diameter and threshold voltages of the CNTFET for various chirality settings. The threshold voltage of the CNTFET for a diameter of 1.487 nm is 0.29 V, so a p-type CNTFET would turn on for inputs of 0 and 1 while an n-type CNTFET will turn on for inputs of 1 and 2.

Throughout this book, we assume the CNTFET channel length (for all simulations) to be 32 nm. Further, the transistor sizes depend on the chirality explicitly specified (and shown in the circuit diagrams appearing in the book). Similar transistor sizes have also been adopted by other authors [11]. In addition, the number of carbon nanotubes is assumed to be one throughout the book. The libraries used for all the simulations follow [9, 10].

Table 2.1 CNT diameter (D_{CNT}) and threshold voltage (V_{th}) for different chirality settings for CNTFET

Chirality	D_{CNT} (nm)	V_{th} (V)
(10, 0)	0.783	0.53
(13, 0)	1.018	0.41
(19, 0)	1.487	0.29

2.3 Ternary Logic and Algebra

Ternary number system permits three-state logic representation. Every ternary number can be represented with a digit from the set {0, 1, 2}. A ternary number with n digits can be expressed as $T_{n-1}T_{n-2} \ldots T_1 T_0$. The equivalent decimal digit can be obtained as given by Eq. (2.5).

$$D = \sum_{i=0}^{n-1} 3^i * T_i \tag{2.5}$$

Equation (2.5) gives the positive decimal number equivalent of the n-digit ternary number. Negative decimal numbers can also be represented using ternary digits and one can convert a ternary string to decimal using Eq. (2.6).

$$D = -3^{n-1} + \sum_{i=0}^{n-2} 3^i * T_i \tag{2.6}$$

The representation of decimal numbers in ternary is given in Table 2.2. A number (in base-10) can be represented more compactly (with fewer digits) in ternary than in binary. For instance, 8_{10} is represented as 1000_2 in binary while it is 22_3 in ternary. Another advantage of the ternary number system is with respect to the interconnections required. Since the number of digits is less, the on-chip and off-chip connections reduce (in comparison to binary). Figure 2.4 depicts the major advantages of ternary logic.

2.3.1 Unary Operators

We start with the familiar binary number system and then discuss extensions to the ternary case. In the binary number system (with base $p = 2$), we have four functions, namely identity ($Y = A$), negation ($Y = \overline{A}$), constant corresponding to $Y = 0$ and constant corresponding to $Y = 1$ as given in Table 2.3 [13] where a_0 and a_1 can

Table 2.2 Representation of decimal numbers in ternary logic system

Decimal	x_2	x_1	x_0
0	0	0	0
1	0	0	1
2	0	0	2
3	0	1	0
4	0	1	1
5	0	1	2
6	0	2	0
−1	1	2	2
−2	1	2	1
−3	1	2	0
−4	1	1	2
−5	1	1	1
−6	1	1	0

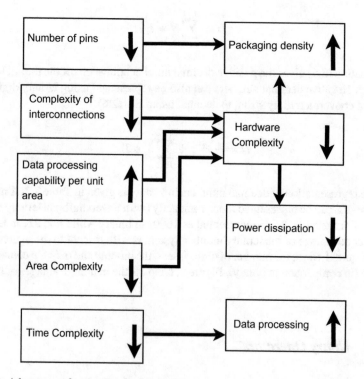

Fig. 2.4 Advantages of ternary Logic circuits [12]

Table 2.3 Unary operators in binary logic system

a_1	a_0	Function	Name
0	0	$Y = 0$	Constant
0	1	$Y = A$	Identity
1	0	$Y = \overline{A}$	Negation
1	1	$Y = 1$	Constant

Table 2.4 Inputs, outputs and operators in binary logic system

Input	Output	Function	Operator
A	0	$Y = 0$	Constant
A	A	$Y = A$	Identity or Buffer
A	$\overline{A}$	$Y = \overline{A}$	Negation or Inverter
A	1	$Y = 1$	Constant

Table 2.5 Some unary operators in ternary logic

Input	Cycle		Inverters			Decisive		
A	A^1	A^2	$\overline{A}$	A_P	A_N	A_0	A_1	A_2
0	1	2	2	2	2	2	0	0
1	2	0	1	2	0	0	2	0
2	0	1	0	0	0	0	0	2

each take values of 0 or 1. Table 2.4 gives an alternative interpretation. In general, a p−base number system allows altogether p^p unary or one-place functions.

In the ternary number system (base $p = 3$), there exists a total of 27 unary operators [13]. Any ternary function can be realized with these unary operators. Table 2.5 shows some of the common unary operators. It is worth noting from Table 2.5 that when $A = 0$, $\overline{A}$, A_P and A_N are 2. However, when $A = 1$, $\overline{A}$, A_P and A_N take different values. $\overline{A}$, A_P and A_N represent three types of inverters in ternary (more discussion is presented in the next chapter). Similarly, we have cycle and decisive operators that play an important role in obtaining low-complexity circuit designs. For example, the decisive operator A_1 is 2 when $A = 1$ and is 0 otherwise.

Table 2.6 gives the entire list of unary operators. We assume that · and + symbols correspond to min (i.e., minimum) and max (i.e, maximum) operators respectively.

A typical entry in Table 2.6 can be verified as follows. Consider, for example, row 2. Y_1 is given by $(0, 0, 1)$ and can be obtained from Y_2. Y_2 itself can be obtained from Y_{24}. In particular, Y_2 given by $\overline{A_P}$, is obtained by what is known as standard ternary inversion of A_P. Then we obtain Y_1 as $max(1, \overline{A_P})$. It turns out that Y_1 plays an important role in the design of an efficient half-adder.

Remark 2.1 It is worth noting that the entries (in the last column of Table 2.6) are not unique. For example, $a_2 = 1$, $a_1 = 1$, $a_0 = 2$ in row 15 can also be expressed

Table 2.6 All possible unary operators in ternary for input A

S.No	a_2	a_1	a_0	Function
1	0	0	0	$Y_0 = 0$
2	0	0	1	$Y_1 = 1 \cdot \overline{A_P}$
3	0	0	2	$Y_2 = A_2$
4	0	1	0	$Y_3 = 1 \cdot A_1$
5	0	1	1	$Y_4 = 1 \cdot \overline{A_N}$
6	0	1	2	$Y_5 = A$
7	0	2	0	$Y_6 = A_1$
8	0	2	1	$Y_7 = \overline{A^2}$
9	0	2	2	$Y_8 = \overline{A_N}$
10	1	0	0	$Y_9 = 1 \cdot A_N$
11	1	0	1	$Y_{10} = 1 \cdot A^2$
12	1	0	2	$Y_{11} = \overline{A^1}$
13	1	1	0	$Y_{12} = 1 \cdot A^1$
14	1	1	1	$Y_{13} = 1$
15	1	1	2	$Y_{14} = 1 + \overline{A_P}$
16	1	2	0	$Y_{15} = A^1$
17	1	2	1	$Y_{16} = 1 + A^1$
18	1	2	2	$Y_{17} = 1 + \overline{A_N}$
19	2	0	0	$Y_{18} = A_N$
20	2	0	1	$Y_{19} = A^2$
21	2	0	2	$Y_{20} = \overline{A_1}$
22	2	1	0	$Y_{21} = \overline{A}$
23	2	1	1	$Y_{22} = 1 + \overline{A}$
24	2	1	2	$Y_{23} = 1 + \overline{A_1}$
25	2	2	0	$Y_{24} = A_P$
26	2	2	1	$Y_{25} = 1 + A_P$
27	2	2	2	$Y_{26} = 2$

as $1 + \overline{A^1}$. Further, A_N in row 19 is the same as decisive operator A_0 while the combination in row 3 also represents $\overline{A_P}$. Many other interesting possibilities exist for the entries in the last column in general.

Having listed the complete set of unary operators in ternary, it is of interest to examine if some operators are related to others. Table 2.7 provides various relationships. These relationships help in minimizing ternary logic expressions and in the development of a synthesis methodology.

The circuit realization of unary operators is discussed in the chapters to follow. We now discuss another aspect that is of interest in ternary function minimization. This is a graphical representation and can be combined with the algebra presented earlier.

Table 2.7 Relationships between unary operators

$A_0 + A_1 = A_P$	$A_1 + A_2 = \overline{A_N}$
$A_N + A_P = A_P$	$A_0 + A_2 = \overline{A_1}$
$A_P + \overline{A_P} = 2$	$A_N + \overline{A_N} = 2$
$A_P + \overline{A_N} = 2$	$A_N + \overline{A_P} = \overline{A_1}$
$A_P + A^1 = A_P$	$A_N + A^1 = A_P$
$A_P + \overline{A} = A_P$	$A_N + \overline{A} = \overline{A}$
$A_P + A^2 = 1 + A_P$	$A_N + A^2 = A^2$
$\overline{A_N} + A^2 = 2$	$\overline{A_P} + A^2 = \overline{A_1}$
$\overline{A} + A^1 = A_P$	$\overline{A_N} + A^1 = 1 + \overline{A_N}$
$A_P \cdot A^1 = A^1$	$A_P \cdot A^2 = A_N$
$A_P \cdot \overline{A} = \overline{A}$	$A_N \cdot A^1 = 1 \cdot A_N$
$A_N \cdot A^2 = A_N$	$A_N \cdot \overline{A} = A_N$
$A \cdot 1 = 1 \cdot \overline{A_N}$	$A \cdot 2 = A$

2.3.2 Graphical Representation

A graphical representation of ternary functions can be obtained using a 'cube-like' structure [14] shown in Fig. 2.5. Layers of the cube correspond to faces or entities at *unit-distance* from the faces. There are 3^3 (or 27) distinct minterms for ternary functions of three ternary variables and these are depicted in Fig. 2.5. The representation is useful for reducing ternary functions of three variables.

Note that, in the ternary case, the output (value of the function) can be 0, 1 or 2 in general. Hence the 'presence' of a term alone is not adequate to completely characterize the function of interest. Therefore, we have labels of 0, 1 and 2 indicated in Fig. 2.5. From the cube, Eq. (2.7) can be obtained.

$$
\begin{aligned}
Y = {} & 1 \cdot c_0 b_1 a_0 + 2 \cdot c_0 b_2 a_0 + 1 \cdot c_1 b_0 a_0 + 2 \cdot c_1 b_0 a_1 \\
& + 2 \cdot c_2 b_0 a_0 + 1 \cdot c_2 b_0 a_2 + 1 \cdot c_2 b_1 a_1 + 1 \cdot c_2 b_2 a_1
\end{aligned} \tag{2.7}
$$

Equation (2.7) can be simplified as follows. Here, for simplification, terms which are equal to '0' such as '$0 \cdot c_0 b_0 a_0$' have been added.

$$
\begin{aligned}
Y = {} & 0 \cdot c_0 b_0 a_0 + 1 \cdot c_0 b_1 a_0 + 2 \cdot c_0 b_2 a_0 \\
& + 1 \cdot c_1 b_0 a_0 + 2 \cdot c_1 b_0 a_1 + 0 \cdot c_1 b_0 a_2 \\
& + 2 \cdot c_2 b_0 a_0 + 0 \cdot c_2 b_0 a_1 + 1 \cdot c_2 b_0 a_2 \\
& + 2 \cdot c_2 b_0 a_0 + 1 \cdot c_2 b_1 a_1 + 1 \cdot c_2 b_2 a_1 \\
Y = {} & c_0 a_0 \cdot (0 \cdot b_0 + 1 \cdot b_1 + 2 \cdot b_2) \\
& + c_1 b_0 \cdot (1 \cdot a_0 + 2 \cdot a_1 + 0 \cdot a_2) \\
& + c_2 b_0 \cdot (2 \cdot a_0 + 0 \cdot a_1 + 1 \cdot a_2)
\end{aligned}
$$

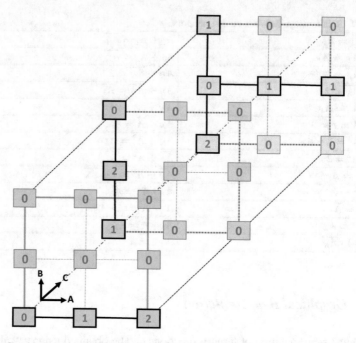

Fig. 2.5 Graphical representation of a three-variable function

$$+c_2 a_1 \cdot (0 \cdot b_0 + 1 \cdot b_1 + 1 \cdot b_2) \tag{2.8}$$

Now, from the unary operator definitions, Eq. (2.8) can be rewritten as Eq. (2.9).

$$
\begin{aligned}
Y &= c_0 a_0 \cdot (B) + c_1 b_0 \cdot (A^1) \\
&\quad + c_2 b_0 \cdot (A^2) \\
&\quad + c_2 a_1 \cdot (1 \cdot \overline{B_N}) \\
Y &= c_0 a_0 \cdot (B) + c_1 b_0 \cdot (A^1) \\
&\quad + c_2 \cdot (b_0 \cdot A^2 + a_1 \cdot (1 \cdot \overline{B_N}))
\end{aligned} \tag{2.9}
$$

2.4 Impact of Process, Voltage and Temperature (PVT) Variations

The variations in the process, voltage and temperature influence, in general, the circuit performance metrics such as delay and power. The process variation pertains to fabrication of the device. In CMOS circuits, the process variation includes the variation in the gate oxide thickness and diffusion depths. This results in the variation of the threshold voltage which in turn results in the switching of the device. Further, variation in the threshold voltage causes changes in current in CMOS. On the other hand, the threshold voltage of the CNTFET depends only on the device geometry and not on the process [1]. Hence, the impact of process variation on the threshold voltage is small in the case of the CNTFET-based devices. While there have been concerns about the nature of CNT (metallic/semi-conducting) grown, techniques have been proposed for removal of metallic CNTs [15]. Another aspect of interest is the effect of the process (of CNT growth) on the diameter. A recent study on the process variation in a multichannel CNTFET [16] reveals that the variations in the diameter of the CNT have less effect in a multichannel CNTFET (in comparison to single channel).

A third aspect of interest is the impact of variation of temperature on the performance of CNTFET-based designs. Temperature variation is unavoidable because of continuous usage and the quest for high performance. In CMOS-based designs, a prominent issue [17, 18] is bias temperature instability. However, thermal stability is higher in a CNTFET [17] when compared with silicon-based devices. The output current degradation (due to self heating) has been observed to be less than 2% [17]. Further, it has been shown [17] that the effect of temperature on the delay and power is very less.

2.5 Summary

In this chapter, we have presented terminology that will be used in the remainder of this book. Select unary operators have interesting consequences in terms of leading to efficient adder designs. We discuss these in subsequent chapters.

References

1. Raychowdhury, A., Roy, K.: Carbon-Nanotube-based voltage-mode multiple-valued logic design. IEEE Trans. Nanotechnol. 4(2), 168–179 (2005)
2. Saito, R., Fujita, M., Dresselhaus, G., Dresselhaus, M.S.: Electronic structure of graphene tubules based on c_{60}. Phys. Rev. B 46, 1804–1811 (1992)
3. Iijima, S.: Helical microtubules of graphitic carbon. Nature 354(6348), 56–58 (1991)
4. Javey, A., Tu, R., Farmer, D.B., Guo, J., Gordon, R.G., Dai, H.: High-performance n-type carbon nanotube field-effect transistors with chemically doped contacts. Nano-Micro Lett. 5(2), 345–348 (2005)
5. Tans, S.J., Verschueren, A.R.M., Dekker, C.: Room-temperature transistor based on a single carbon nanotube. Nature 393, 49–52 (1998)
6. Rahman, A., Guo, J., Datta, S., Lundstrom, M.S.: Theory of ballistic nanotransistors. IEEE Trans. Electron Devices 50(10), 1853–1864 (2003)
7. Heinze, S., Tersoff, J., Martel, R., Derycke, V., Appenzeller, J., Avouris, Ph.: Carbon nanotubes as Schottky barrier transistors. Phys. Rev. Lett. 89(10), 106801-1–106801-4 (2002)

8. Najari, M., Fregonese, S., Maneux, C., Mnif, H., Masmoudi, N., Zimmer, T.: Schottky barrier carbon nanotube transistor: compact modeling, scaling study, and circuit design applications. IEEE Trans. Electron Devices **58**(1), 195–205 (2011)
9. Deng, J., Wong, H.-S.P.: A compact SPICE model for carbon-nanotube field-effect transistors including nonidealities and its application - Part I: Model of the Intrinsic Channel Region. IEEE Trans. Electron Devices **54**(12), 3186–3194 (2007)
10. Deng, J., Wong, H.S.P.: A compact SPICE model for carbon-nanotube field-effect transistors including nonidealities and its application - Part II: Full Device Model and Circuit Performance Benchmarking. IEEE Trans. Electron Devices **54**(12), 3195–3205 (2007)
11. Lin, S., Kim, Y.B., Lombardi, F.: CNTFET-based design of ternary logic gates and arithmetic circuits. IEEE Trans. Nanotechnol. **10**(2), 217–225 (2011)
12. Kameyama, M., Kawahito, S., Higuchi, T.: A multiplier chip with multiple-valued bidirectional current-mode logic circuits. IEEE Computers **23**(4), 43–56 (1988)
13. Miller, D.M., Thornton, M.A.: Multiple Valued Logic: Concepts and Representations. Morgan and Claypool Publishers (2008)
14. Hurst, S.L.: An extension of binary minimization techniques to ternary equations. Comput. J. **11**(3), 277–286 (1968)
15. Krupke, R., Heinrich, F., Lohneysen, H.V., Kappes, M.M.: Separation of metallic from semiconducting single-walled carbon nanotubes. Science **301**(5631), 344–347 (2003)
16. Raychowdhury, A., De, V.K., Borkar, S.Y., Roy, K., Keshavarzi, A.: Variation tolerance in a multichannel carbon-nanotube transistor for high-speed digital circuits. IEEE Trans. Electron Devices **56**(3), 383–392 (2009)
17. Xing, C.J., Yin, W.Y., Liu, L., Huang, J.: Investigation on self-heating effect in carbon nanotube field-effect transistors. IEEE Trans. Electron Devices **58**(2), 523–529 (2011)
18. Kim, T.H., Persaud, R., Kim, C.H.: Silicon odometer: an on-chip reliability monitor for measuring frequency degradation of digital circuits. IEEE J. Solid-State Circuits **43**(4), 874–880 (2008)

Chapter 3
CNTFET-Based Circuits for Basic Logic Elements

In this chapter, we begin the transistor-based design study with basic logic elements. We start with ternary inversion and present CNTFET-based circuits. We use transistors (CNTFETs) as load analogous to [1]. However, our approach in general, does not employ decoder and encoder pairs (unlike [1]).

3.1 The Inverter

In binary logic, the notion of an inverter is unique. It is expressed by Eq. (3.1).

$$\overline{A} = 1 - A \tag{3.1}$$

Given Eq. (3.1), we note that $\overline{A}$ is 1 when $A = 0$ (similarly, $\overline{A}$ is 0 when $A = 1$). However, in ternary (and beyond), the inverter can take "many forms". We begin with the discussion of an extension of the binary inverter called as the Standard Ternary Inverter (denoted by STI).

3.1.1 Standard Ternary Inverter

Equation (3.1) readily generalizes to yield the description for what is known as the standard ternary inverter. In particular, we have Eq. (3.2) [2].

$$\overline{A} = 2 - A \tag{3.2}$$

From Eq. (3.2), we obtain the truth table for a standard ternary inverter. It is given by Table 3.1

© Springer Nature Switzerland AG 2020
K. Sridharan et al., *Low-Complexity Arithmetic Circuit Design in Carbon Nanotube Field Effect Transistor Technology*, Carbon Nanostructures,
https://doi.org/10.1007/978-3-030-50699-5_3

Table 3.1 Truth table for standard ternary inverter

A	$\overline{A}$
0	2
1	1
2	0

Table 3.2 Truth table for positive ternary inverter

A	A_P
0	2
1	2
2	0

Fig. 3.1 CNTFET-based realization of a Standard Ternary Inverter (STI)

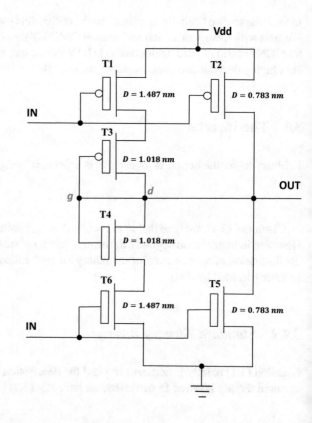

It is worth noting from Table 3.2 that the "middle" (input) level (namely, level 1) remains unchanged at the output. The CNTFET-based realization of an STI is shown in Fig. 3.1.

The circuit shown in Fig. 3.1 operates as follows. Let Vdd be 0.9V. A p-type CNTFET will be ON when the (given) gate to source voltage is less than the threshold voltage. In this case, the CNTFET $T2$ with a diameter of 0.783 nm will have a threshold voltage of -0.53 V while the CNTFET $T1$ with a diameter of 1.487 nm will have a threshold voltage of -0.29 V. Suppose the input IN given to the circuit is 0 V (i.e., logic '0'). Both $T1$ and $T2$ will be ON since the gate to source voltage difference is -0.9 V (and this is lesser than the threshold voltage of the CNTFETs $T1$ and $T2$). The other p-type CNTFET $T3$ is a diode-connected transistor (since the drain of $T3$ is connected to gate of $T3$). CNTFET $T3$ will be ON, only if the diode connected n-type CNTFET (drain of $T4$ is connected to gate of $T4$) is ON (note that gate of $T3$ and $T4$ are connected; also drain of $T3$ and $T4$ are connected). Since the CNTFETs $T4$ and $T6$ are OFF, CNTFET $T3$ is OFF. Hence, there is no path to the output through the CNTFETs $T1$ and $T3$. However, there is a path to output from V_{dd} through $T2$, leading to an output voltage of 0.9 V, (namely, logic '2'). Note that $T4$, $T5$ and $T6$ are OFF when $IN = 0V$.

Now consider another possibility for IN, namely 0.9 V (namely, logic '2'). An n-type CNTFET will be ON, when the gate to source voltage is greater than the threshold voltage. Since CNTFETs $T5$ and $T6$ have diameters of 0.783 nm and 1.487 nm respectively and these correspond to threshold voltages of 0.53 V and 0.29 V respectively (for $T5$ and $T6$), $T5$ and $T6$ will be ON. Since the CNTFET $T4$ is diode connected, this CNTFET will be ON only if the corresponding p-type CNTFET is ON. Note that $T1$, $T2$ and $T3$ are OFF here. Hence, CNTFET $T4$ is OFF and there is no path to output via $T4$. However, there is a path from output to ground via CNTFET $T5$, (which is ON) leading to output of 0 V (corresponding to logic '0').

We now consider the third possibility for IN, namely 0.45 V (logic '1'). Now the CNTFETs $T1$, $T3$, $T4$ and $T6$ are ON while the remaining two CNTFETs ($T2$ and $T5$) are OFF. This leads to a path from V_{dd} to ground. Since the circuit behaves as a voltage divider circuit, the output voltage will be 0.45 V ($\frac{V_{dd}}{2}$), which corresponds to logic '1'. This validates Table 3.1 for the STI.

3.1.2 Positive Ternary Inverter

Accomplishing inversion of the two "extremes" (namely, level 0 and level 2) is possible with lower design complexity in the ternary case as pointed out in [2]. This is of direct interest in CNTFET-based circuits where multiple threshold voltages are achievable by varying the diameter of the carbon nanotube. One such configuration is called as the Positive Ternary Inverter (denoted by PTI). The PTI is expressed by Eq. (3.3) [2].

$$\bar{x} = \begin{cases} 2, & \text{if } x \neq 2 \\ 0 & \text{if } x = 2 \end{cases} \tag{3.3}$$

Fig. 3.2 CNTFET-based
realization of a Positive
Ternary Inverter (PTI)

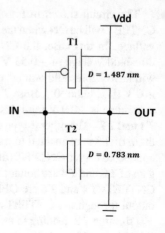

The truth table for a positive ternary inverter is given by Table 3.2. The CNTFET-based realization of a PTI is shown in Fig. 3.2.

Remark 3.1 Comparing Table 3.2 with Table 3.1, we observe that input levels of 0 and 2 map to 2 and 0 respectively (as in STI). However, an input level of 1 results in an output level of 2 in PTI.

A CNTFET-based *positive ternary inverter* shown in Fig. 3.2 operates as follows. Since the threshold voltage of p-type CNTFET $T1$ is -0.29 V, it will be ON for inputs corresponding to logic '0' and '1' while it will be OFF for logic '2'. Similarly, the n-type CNTFET $T2$ has a threshold voltage of 0.53 V and hence it will be OFF for both the inputs corresponding to logic '0' and '1' while it will be ON for (input) logic '2'. Suppose the input voltage IN is 0 V (logic '0'). CNTFETs $T1$ will be ON while $T2$ will be OFF therefore the output OUT is 0.9 V (Vdd; logic '2').

Now suppose the given input IN is 0.45 V (namely, logic '1'). CNTFET $T1$ will have a gate to source voltage of -0.45 V which is lesser than the threshold voltage, so $T1$ will be ON. However, the gate to source voltage of n-type CNTFET $T2$ is 0.45 V, which is less than its threshold voltage 0.53 V. Hence, the CNTFET $T2$ will be OFF. As a result, output OUT corresponds to logic '2' which is 0.9 V (V_{DD}).

Finally, consider an input voltage IN of 0.9 V (namely, logic '2'). Now, CNTFET $T1$ will be OFF because the gate to source voltage is 0 V, which is higher than the threshold voltage. However, CNTFET $T2$ will be ON since its gate to source voltage is 0.9 V, which is higher than the threshold voltage 0.53 V. Hence, output voltage OUT is 0 V due to a path from output to ground. This validates Table 3.2.

It is worth noting that the output of a PTI is a two-valued signal, while the output of the STI is a three-valued signal. Hence, the output of the PTI, A_P, can be used as a control signal to switch any pair of CNTFETs.

Table 3.3 Truth table for negative ternary inverter

A	A_N
0	2
1	0
2	0

Fig. 3.3 CNTFET-based realization of a Negative Ternary Inverter (NTI)

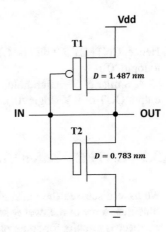

3.1.3 Negative Ternary Inverter

Analogous to PTI, a variation called as the Negative Ternary Inverter (denoted by NTI), is also possible in the ternary setting. This is expressed by Eq. (3.4).

$$\overline{x} = \begin{cases} 2, & \text{if } x = 0 \\ 0 & \text{if } x \neq 0 \end{cases} \tag{3.4}$$

Based on Eq. (3.4), the truth table for a negative ternary inverter is given by Table 3.3. The CNTFET-based realization of an NTI is shown in Fig. 3.3.

Remark 3.2 As seen for PTI, inversion of level 2 and level 0 happen for NTI as well and it turns out that this happens with lower design complexity than STI.

The *CNTFET-based negative ternary inverter* in Fig. 3.3 operates as follows. Since the threshold voltage of the p-type CNTFET $T1$ is -0.53 V, it will be ON for input corresponding to logic '0', while it will be OFF for inputs corresponding to logic '1' and logic '2'. Similarly, the n-type CNTFET $T2$ has a threshold voltage of 0.29 V and hence it will be OFF for input corresponding to logic '0' while it will be ON for inputs corresponding to logic '1' and logic '2'.

Now suppose the input IN is 0.45 V (logic '1'). Since this implies that the CNTFET $T1$ will have a gate to source voltage of 0.45 V which is higher than the threshold voltage, CNTFET $T1$ will be OFF. However, the gate to source voltage of n-type CNTFET $T2$ is 0.45 V which is higher than the threshold voltage (namely, 0.29 V).

Table 3.4 Additional unary operators for ternary logic

Input	Cycle		Decisive		
A	A^1	A^2	A_0	A_1	A_2
0	1	2	2	0	0
1	2	0	0	2	0
2	0	1	0	0	2

Hence, CNTFET $T2$ will be ON. As a result, the output OUT is 0 V(corresponding to logic '0').

A similar explanation holds when the input IN is 0.9 V (logic '2'), leading to an output OUT of 0 V (logic '0').

3.2 CNTFET-Based Circuits for Other Unary Operators

We have discussed the realization of three "types" of inverters. STI, PTI and NTI constitute three of the twenty seven unary operators of ternary logic. In this section, we discuss circuits for some other unary operators which also play a valuable role in low-complexity arithmetic circuit design.

Table 3.4 lists a few unary operators. These include two *cycle operators*, namely A^1 and A^2, and three *decisive operators*, namely A_0, A_1 and A_2. These operators also help in designing low-complexity arithmetic circuits. The CNTFET-based realization of the cycle operators is shown in Figs. 3.4 and 3.5. Figures 3.6, 3.7 and 3.8 show the CNTFET-based realization of the decisive operators. We describe the operation of some of these circuits next.

3.2.1 Cycle Operator, A^1

In a cycle operator, the input to output relationship is similar to rotation of digits. Table 3.4 lists the two cycle operators A^1 and A^2. A^1 is rotation of A 'once' while A^2 is rotation of A 'twice'.

Figure 3.4 shows the CNTFET-based circuit for A^1. The circuit operates as follows. The CNTFET $T1$ has a threshold voltage of -0.29 V so this will be ON for inputs corresponding to logic '0' and '1' and OFF for logic '2'. Similarly, the n-type CNTFET $T2$ is ON only for logic '2', since the threshold voltage of $T2$ is 0.53 V. The third CNTFET $T3$ is n-type and will be ON for logic '0', since the input of $T3$ is connected to A_N (and this will be high only for $A = 0$). Now suppose the input to the circuit given in Fig. 3.4 is logic '0' (i.e., 0 V). This results in CNTFETs $T1$ and

Fig. 3.4 CNTFET-based circuit for cycle operator A^1

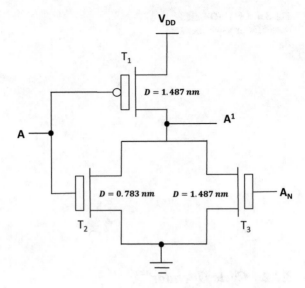

Fig. 3.5 CNTFET-based circuit for cycle operator A^2

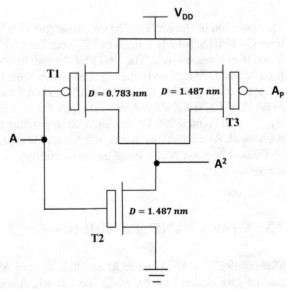

$T3$ being ON while the CNTFET $T2$ is OFF. The circuit acts as a voltage divider resulting in an output of 0.45 V (corresponding to logic '1').

Suppose the input corresponds to logic '1' (i.e., 0.45 V). For this input, only the CNTFET $T1$ is ON while the remaining two CNTFETs are OFF. This causes the output to 'connect to V_{DD}, which is logic '2'. Finally, when the input 'A' is 0.9 V (i.e., logic '2'), the CNTFET $T2$ is ON while all other CNTFETs are OFF leading to 'OUT' getting connected to ground. The circuit output is therefore 0 V.

Fig. 3.6 CNTFET-based
circuit for decisive operator
A_0

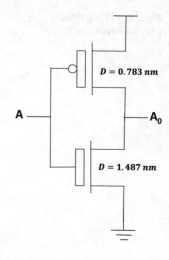

3.2.2   Cycle Operator, A^2

The operation of the *CNTFET-based circuit for A^2* is as follows. Figure 3.5 shows
three CNTFETS, namely $T1$, $T2$ and $T3$ with threshold voltages of -0.53 V, -0.29
V and 0.29 V respectively. The CNTFET $T1$ will be ON for input corresponding to
logic '0' and OFF for (any) other input. Similarly, the n-type CNTFET $T2$ will be ON
for inputs corresponding to logic '1' and '2' and OFF for input logic '0'. With respect
to the third CNTFET $T3$ (which is *p*-type), we note that the input is connected to
A_P. $T3$ will be therefore ON for input corresponding to logic '2' and OFF for any
other input. As a result, the output of the circuit is described by Table 3.4.

We next discuss CNTET-based circuits for ternary NAND and ernary NOR. These
are based on [1, 2].

3.3   Ternary NAND and NOR Gates

Mathematically, NAND and NOR are derived from AND and OR respectively so
we first give expressions for AND and OR which apply to multivalued logic. In
particular, Eq. (3.5) describes AND of n inputs given by $x_1, x_2, \ldots, x_n$.

$$AND(x_1, x_2, \ldots, x_n) = \min(x_1, x_2, \ldots, x_n) \tag{3.5}$$

It is worth noting that when $x_i, i = 1, \ldots, n$ are binary valued, LHS of Eq. (3.5)
will evaluate to 1 if and only if $x_1 = x_2 = \cdots = x_n = 1$ (and will be 0 otherwise). In
the ternary case, AND will evaluate to 0 if at least one x_i is 0; also, the result would
be 1 if none of the x_i are 0 and at least one $x_i = 1$; when all the x_i are 2, AND would
evaluate to this value as well.

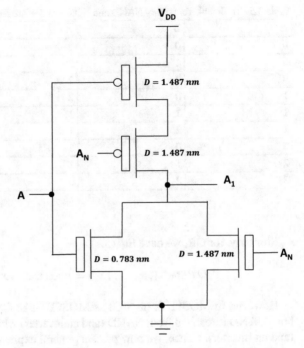

Fig. 3.7 CNTFET-based circuit for decisive operator A_1

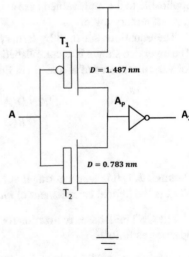

Fig. 3.8 CNTFET-based circuit for decisive operator A_2

Table 3.5 Truth table for ternary NAND and NOR; · and + are ternary AND and OR respectively

A	B	$\overline{A \cdot B}$	$\overline{A + B}$
0	0	2	2
0	1	2	1
0	2	2	0
1	0	2	1
1	1	1	1
1	2	1	0
2	0	2	0
2	1	1	0
2	2	0	0

Similarly, for OR, we have Eq. (3.6).

$$OR(x_0, x_1, x_2, \ldots, x_n) = \max(x_0, x_1, x_2, \ldots, x_n) \tag{3.6}$$

However, in MOSFETs (as well as MOSFET-like CNTFETs of interest in this book), AND is realized from NAND (and an inverter) while OR is realized via NOR (and an inverter). Hence, we now present general expressions for NAND and NOR applicable to the multivalued case and follow it up with CNTFET-based circuits that support ternary logic.

The equations satisfied by ternary NAND and ternary NOR are given by (3.7). The inversion shown is accomplished by a standard ternary inverter(STI). The output for various choices of A and B (in Eq. (3.7)) is given in Table 3.5.

$$NAND(A, B) = \overline{AND(A, B)}$$
$$NOR(A, B) = \overline{OR(A, B)} \tag{3.7}$$

Remark 3.3 It is worth noting that NAND is the logical complement of min while NOR is the logical complement of max.

The CNTFET-based realization for ternary NAND is shown in Fig. 3.9. The circuit operates as follows.

- Let inputs A and B be 0.9 V (i.e., logic '2'). The n-type CNTFETs $T7$ and $T9$ have a threshold voltage of 0.53 V. Hence, both of these CNTFETs are ON. However, the two n-type CNTFETs $T8$ and $T10$ are OFF because of the diode connected CNTFET $T6$ (which is OFF). It is worth noting that eventhough the gate to source voltage is higher than the threshold voltage (which is 0.29 V), $T8$ and $T10$ are OFF. All the p-type CNTFETs are OFF, since the gate to source voltage is 0 V while the threshold voltages are -0.29 V and -0.53 V. As a result, there is only a path from output to ground, through $T7$ and $T9$. Hence, OUT will be 0 V (i.e., logic '0').

- Now consider the following inputs: A = 0.45 V (logic '1') and B= 0.9 V (logic '2'). Here, the p-type CNTFET $T1$ has a threshold voltage of -0.29 V while the gate source voltage is -0.45 V. Hence, the CNTFET $T1$ is ON. Similarly, CNTFET $T2$ has a threshold voltage of -0.29 V while the gate to source voltage is 0 V causing $T2$ to be OFF. The diode connected CNTFET $T5$ is ON, since the n-type CNTFETs $T6$, $T8$ and $T10$ are ON. The n-type CNTFETs $T8$ and $T10$ have a threshold voltage of 0.29 V, and these will be ON for a (given) input voltage corresponding to logic '1' (and logic '2'). Remaining CNTFETs are OFF (due to similar reasoning based on what the threshold voltage is in relation to the inputs). As a result, there is a path from V_{dd} to ground, through $T1$, $T5$, $T6$, $T8$ and $T10$. This leads to an output (OUT) of 0.45 V ($\frac{V_{DD}}{2}$) corresponding to logic '1'.
- Finally, consider the following inputs: A = 0 V (logic '0') and B = 0 V (logic '0'). In this case, the p-type CNTFETs $T3$ and $T4$ have a threshold voltage of -0.53 V. Gate to source voltage of these CNTFETs are -0.9V, which is lesser than the threshold voltage. Hence, the two CNTFETs $T3$ and $T4$ are ON. The other p-type CNTFETs are OFF because of their threshold voltage. All n-type CNTFETs are OFF since an n-type CNTFET will be ON only when the gate to source voltage is higher than the threshold voltage. In this case, the given input (0 V) is lesser than the threshold voltage (0.29 and 0.53 V). As a result, there is just a path from V_{dd} to the output, through $T3$ and $T4$, leading to OUT = 0.9 V (V_{dd}; $logic$ '2').

Similarly, the CNTFET-based realization of ternary NOR is depicted in Fig. 3.10. The working of the ternary NOR can be explained as follows.

- Let inputs A and B be 0.9V (i.e., logic '2'). The n-type CNTFETs $T9$ and $T10$ have a threshold voltage of 0.53 V leading to turning ON of both the CNTFETs. However, the n-type CNTFETs $T7$ and $T8$ are OFF because of the diode connected CNTFET $T6$ is OFF. Once again, it is worth noting that $T7$ and $T8$ are OFF eventhough the gate to source voltage is higher than the threshold voltage (which is 0.29 V). All the p-type CNTFETs are OFF since the gate to source voltage is 0 V while the threshold voltages are -0.29 V (and -0.53 V). As a result, there is only a path from output to ground, through $T7$ and $T9$. Hence, OUT will be 0 V (i.e., logic '0').
- Now consider the following inputs: A = 0 V (logic '0') and B = 0.45 V (logic '1'). The p-type CNTFET $T3$ and $T4$ have a threshold voltage of -0.53 V. These CNTFETs will be ON for input voltage corresponding to logic '0' alone. Now, $T3$ is ON while $T4$ is OFF. The other p-type CNTFETs $T1$ and $T2$ have a threshold voltage of -0.29 V and these will be ON for both inputs corresponding to logic '0' and '1'. Hence, $T1$ and $T2$ are ON. The n-type CNTFETs $T7$ and $T8$ have a threshold voltage of 0.29 V. These CNTFETs will be ON for inputs corresponding to logic '1' and '2'. For the given inputs, the gate to source voltage of $T7$ is 0 V, which is lesser than the threshold voltage of $T7$ leading to $T7$ being OFF. CNTFET $T8$ has a gate to source voltage of 0.45 V, which is higher than the threshold voltage. Hence, the CNTFET $T8$ is ON. As a result, the diode connected pair $T5$ and $T6$

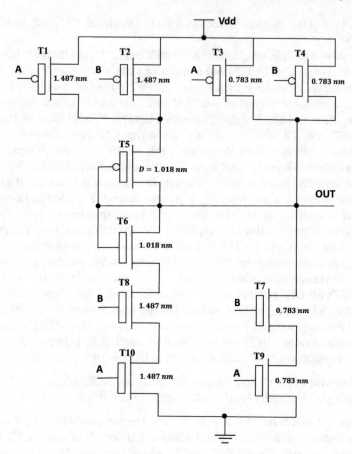

Fig. 3.9 Ternary NAND gate using CNTFET

are ON. The other n-type CNTFETs $T9$ and $T10$ are OFF. There will be only one
path from V_{DD} to ground through the CNTFETs $T1$, $T2$, $T5$, $T6$ and $T8$. Hence,
OUT will be 1.

- Finally, consider the following inputs: A= 0 V (logic '0') and B = 0 V (logic '0').
  In this case, the p-type CNTFETs $T3$ and $T4$ have a threshold voltage of -0.53
  V. Gate to source voltage of these CNTFETs are -0.9V, which is lesser than the
  threshold voltage leading to the two CNTFETs $T3$ and $T4$ being ON. However,
  the other p-type CNTFETs are OFF because of their threshold voltage. All n-type
  CNTFETs are OFF since the n-type CNTFET will be ON only when the gate to
  source voltage is higher than the threshold voltage. Here, the given input is 0 V,
  which is lesser than the threshold voltage (0.29 and 0.53 V). As a result, there is
  just a path from V_{dd} to the output, through $T3$ and $T4$, leading to OUT = 0.9 V
  (V_{dd}; $logic$ '2').

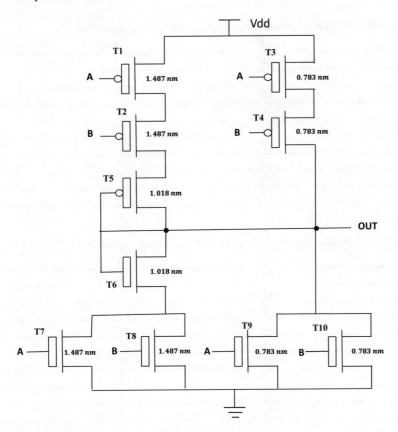

Fig. 3.10 Ternary NOR gate using CNTFET

3.4 Ternary AND and OR Gates

Having described inverters and ternary NAND, we are now in a position to obtain ternary AND. The ternary OR is obtained similarly (from the combination of ternary NOR and inverters). In particular, the STI is used for inversion.

As indicated earlier, the AND function is expressed by Eq. (3.5). For the ternary case, AND is expressed by Table 3.6.

Similarly, ternary OR is expressed by Eq. (3.6) and is given by Table 3.7.

The circuit realization for ternary AND is obtained by combining those for NAND and STI. Similarly, the circuit for ternary OR is obtained by combining those for NOR and STI. This is indicated in Fig. 3.11.

Table 3.6 Truth table for ternary AND where · denotes a ternary operation

A	B	$A \cdot B$
0	0	0
0	1	0
0	2	0
1	0	0
1	1	1
1	2	1
2	0	0
2	1	1
2	2	2

Table 3.7 Truth table for ternary OR where + denotes a ternary operation

A	B	$A + B$
0	0	0
0	1	1
0	2	2
1	0	1
1	1	1
1	2	2
2	0	2
2	1	2
2	2	2

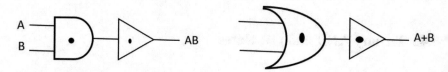

Fig. 3.11 Ternary AND and OR gate realizations

3.5 Ternary D Flip-Flop

A ternary D flip-flop is a sequential memory element to store the data. The clock is a binary clock $(0 - 2 - 0)$ and the inverted clock can be generated using a binary inverter. The D flip-flop is depicted in Fig. 3.12.

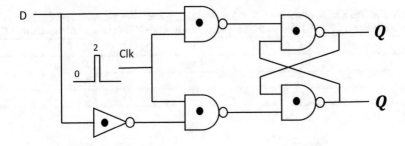

Fig. 3.12 Ternary D flip-flop

Fig. 3.13 CNTFET in
transmission gate
configuration

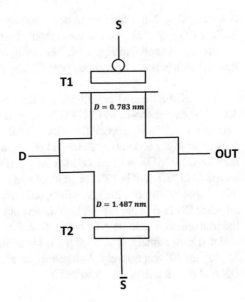

3.6 Ternary Multiplexer

We now discuss an extension of the binary multiplexer to a ternary multiplexer. A
ternary multiplexer serves as a universal building block for realizing various arithmetic circuits in CNTFET technology.

Two types of ternary multiplexers are used in this book for realizing arithmetic
circuits. These correspond to a 3×1 multiplexer and a 2×1 multiplexer. Before
proceeding with the design of the multiplexer, we introduce the "transmission gate
configuration" of a CNTFET. A CNTFET in transmission gate configuration is shown
in Fig. 3.13.

Table 3.8 Truth table of a ternary 3×1 multiplexer; p and n are the gate control signals to the corresponding p-CNTFET and n-CNTFET respectively while Y is the output of the ternary MUX

S	p	n	Y
0	S	S_N	D_0
1	$\overline{S_1}$	S_1	D_1
2	S_P	$\overline{S_P}$	D_2

3.6.1   CNTFET-Based Ternary 3 × 1 Multiplexer

A ternary 3×1 multiplexer is described by Table 3.8, where S is the select line while D_0, D_1 and D_2 are the data (input) lines. The CNTFET-based circuit for the multiplexer is shown in Fig. 3.14. The multiplexer is designed using the CNTFET in transmission gate configuration controlled by the unary operators of the select signal.

The CNTFET-based circuit for the ternary 3×1 multiplexer is shown in Fig. 3.14. The circuit shows 14 CNTFETs and two binary inverters. The operation of this circuit is as follows. Suppose the select signal, S, to the TMUX is logic '0' (i.e., 0 V). This leads to S_N becoming '2' while both S_1 and S_P become '0'. A p-type CNTFET used in the TMUX will be ON for an input corresponding to logic '0', while an n-type CNTFET will be ON for an input corresponding to logic '2' (and OFF for any other input value). For the given S=0, only the CNTFETs $T11$ and $T12$ are ON and all other CNTFETs are OFF. This causes data $D1$ to be passed to output Y, through the transmission gate pair $T11$ and $T12$. Now consider another possibility for the select signal, namely, S ='2'. This leads control signals S_N, S_1 and S_P to become '2', '0' and '0' respectively. As a result, the transmission gate pair $T13$ and $T14$ turn ON and pass the data $D2$ to output Y.

3.6.2   CNTFET-Based Ternary 2 × 1 Multiplexer

Generally, it is assumed that the 'minimum' size multiplexer that is appropriate for the ternary setting is a 3×1 multiplexer with three data lines and one select signal. In particular, the select signal is assumed to be three-valued signal. However, in some cases (for example, certain arithmetic circuits), the select signal may have just two 'defined' logic levels (i.e., (0, 1), (1, 2) or (0, 2)) instead of three logic levels. In this scenario, the use of the conventional ternary 3×1 multiplexer will result in additional hardware (and higher transistor count). Hence, a new 2×1 ternary multiplexer is proposed which has just two data lines and one select signal (that is two-valued). Table 3.9 gives the description of such a ternary 2×1 multiplexer with the corresponding control signal to the transmission gate. The CNTFET-based

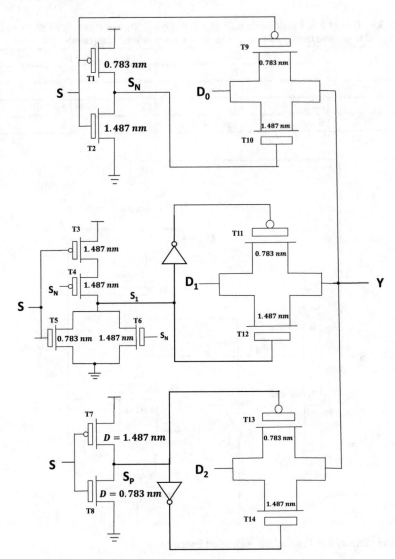

Fig. 3.14 CNTFET-based 3 × 1 ternary multiplexer

realizations for these multiplexers (few variations are possible as indicated) are given in Figs. 3.15, 3.16 and 3.17.

The operation of the ternary 2 × 1 multiplexer is as follows. Let us consider the circuit shown in Fig. 3.15. This circuit has two inputs, D_0 and D_1, with one select signal S. Suppose the input for the select signal S is 0 V (i.e., logic '0'). This leads to CNTFETs $T3$ and $T4$ becoming ON because of the control signal S_N (becoming '2'). All other CNTFETs are OFF and this leads the output Y to connect to the input

Table 3.9 Ternary 2×1 multiplexer with data lines D_0 and D_1 and select signal S; S_P and S_N correspond to positive ternary inverter and negative ternary inverter respectively

S	D_0		D_1	
	P	N	P	N
0, 1	S	S_N	S_N	$\overline{S_N}$
1, 2	$\overline{S_P}$	S_P	S_P	$\overline{S_N}$
0, 2	S	S_N	S_N	S

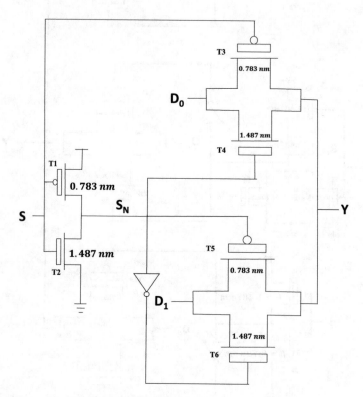

Fig. 3.15 Ternary 2×1 multiplexer with select lines 0 and 1

D_0 via the transmission gate $T3$ and $T4$. Similar explanation holds for other circuits for the ternary 2×1 multiplexer (in Figs. 3.16 and 3.17).

Remark 3.4 The ternary 2×1 multiplexer discussed requires 8 CNTFETs while the conventional ternary 3×1 multiplexer takes 16 CNTFETs.

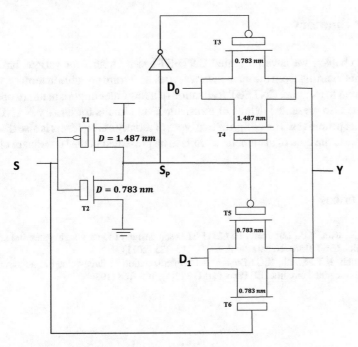

Fig. 3.16 Ternary 2 × 1 multiplexer with select lines 1 and 2

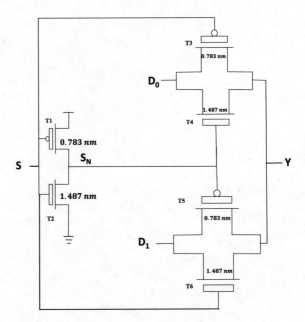

Fig. 3.17 Ternary 2 × 1 multiplexer with select lines 0 and 2

3.7   Summary

In this chapter, we have presented CNTFET-based circuits for various basic logic elements. Starting from the three inverter types which are possible in ternary, we have proceeded to examine CNTFET realization of a few other important unary operators. We have also presented details of transistor-level circuits for ternary NAND, NOR and other primitives. The design of two types of ternary multiplexers is also discussed. These form the core of arithmetic circuits to be presented in the subsequent chapters.

References

1. Lin, S., Kim, Y.B., Lombardi, F.: CNTFET-based design of ternary logic gates and arithmetic circuits. IEEE Trans. Nanotechnol. **10**(2), 217–225 (2011)
2. Mouftah, H.T., Smith, K.C.: Design and implementation of three-valued logic systems with MOS integrated circuits. IEE Proc. Part G **127**(4), 165–168 (1980)

Chapter 4
CNTFET-Based Design of a Single Ternary Digit Adder

4.1 Ternary Addition—Basic Ideas

Just as we add two bits in a (binary) half-adder and three bits in a (binary) full-adder, one can consider addition of two or three ternary digits. The addition of two ternary digits is expressed by Table 4.1. The extension to three ternary digits is presented in Table 4.2.

Remark 4.1 Some typical entries in Table 4.1 can be verified as follows: Let A and B be given by 2 and 1 respectively. Adding these two gives 3 (in base-10) which can be expressed as $1 \times 3^1 + 0 \times 3^0$ hence, the SUM (trit) is 0 while the CARRY (trit) is 1. Similarly, when A and B are given by 2 and 2 respectively, their addition gives 4 (in base-10) which is expressible as $1 \times 3^1 + 1 \times 3^0$ hence, the SUM (trit) is 1 while the CARRY (trit) is 1.

Similarly, select entries in Table 4.2 can be verified as follows. Consider $A = 1, B = 2, C = 1$. The addition of these gives 4 (in base 10) which can be expressed as $1 \times 3^1 + 1 \times 3^0$, hence SUM = 1, CARRY = 1. Further, when $A = B = C = 2$, the addition leads to 6 (base 10) which can be written as $2 \times 3^1 + 0 \times 3^0$, hence SUM = 0, CARRY = 2.

4.2 Review of Ternary Multiplexer and Key Unary Operators

It is well-known that one approach to combinational logic design in the binary setting is via multiplexers. The same can be extended to ternary logic as well. By careful design of CNTFET-based ternary multiplexers as well as judicious choice of input for the ternary multiplexers, one can obtain low-complexity designs of arithmetic circuits. We therefore review these two components here.

© Springer Nature Switzerland AG 2020

K. Sridharan et al., *Low-Complexity Arithmetic Circuit Design in Carbon Nanotube Field Effect Transistor Technology*, Carbon Nanostructures,
https://doi.org/10.1007/978-3-030-50699-5_4

Table 4.1 Addition of two ternary digits, A and B

A	B	SUM	CARRY
0	0	0	0
0	1	1	0
0	2	2	0
1	0	1	0
1	1	2	0
1	2	0	1
2	0	2	0
2	1	0	1
2	2	1	1

A ternary 3×1 multiplexer is described by Table 4.3, where S is the select line while D_0, D_1 and D_2 are the data lines. The CNTFET-based circuit for the multiplexer is shown in Fig. 4.1. The multiplexer is designed using the CNTFET in transmission gate configuration and controlled by the unary operators of the select signal.

Inputs for the ternary multiplexer are obtained from the list of unary operators presented earlier (and some variations). Chapter 2 presents various unary operators (in ternary) for an input A. We recall that the unary operators include the three forms of inverters, cycle operators as well as decisive operators. The basic unary operators are listed in Table 4.4.

4.3 Direct Realization of Ternary Half and Full Adders

In order to obtain a direct realization, we need to interpret the truth tables. In other words, one needs to obtain an algebraic expression for the sum and carry outputs from the truth tables. Towards this end, we first discuss the generation of the SUM output for a binary half-adder. Let A and B be two binary numbers which are input to the half-adder. The SUM output can then be written in terms of unary operators, $\overline{A}$ and $\overline{B}$ as $0 \cdot (\overline{A}\ \overline{B}) + 1 \cdot (\overline{A}B) + 1 \cdot (A\overline{B}) + 0 \cdot (AB)$ where $\overline{A}B$ corresponds to binary AND of $\overline{A}$ and B (similar interpretation holds for other combinations). As indicated earlier, the dot ($\cdot$) symbol corresponds to "min" (i.e., minimum) while the plus (+) symbol corresponds to "max" (i.e., maximum). Since the min of 0 and $(\overline{AB})$ yields 0, terms involving "AND" with 0 can be omitted.

We can now extend this to the ternary case. In place of $\overline{A}$, we have a different set of unary operators. In particular, we have A_0, A_1 and A_2. When $A = 0$, $A_0 = 2$. Similarly, when $A = 1$, $A_1 = 2$. Also, $A_2 = 2$ corresponds to $A = 2$. $A_2 B_0$, $B_0 A_1$ etc. can be obtained via AND gates (ternary AND gates are suggested in [1], however we can obtain them more efficiently via binary AND gates since each of these inputs has only two levels [2]). Similarly, + is accomplished via a binary OR gate [2]. The

Table 4.2 Addition of three ternary digits, A, B and C

A	B	C	SUM	CARRY
0	0	0	0	0
0	0	1	1	0
0	0	2	2	0
0	1	0	1	0
0	1	1	2	0
0	1	2	0	1
0	2	0	2	0
0	2	1	0	1
0	2	2	1	1
1	0	0	1	0
1	0	1	2	0
1	0	2	0	1
1	1	0	2	0
1	1	1	0	1
1	1	2	1	1
1	2	0	0	1
1	2	1	1	1
1	2	2	2	1
2	0	0	2	0
2	0	1	0	1
2	0	2	1	1
2	1	0	0	1
2	1	1	1	1
2	1	2	2	1
2	2	0	1	1
2	2	1	2	1
2	2	2	0	2

Table 4.3 Truth table of a ternary 3×1 multiplexer; p and n are the gate control signals to the corresponding p-CNTFET and n-CNTFET respectively

S	p	n	Y
0	S	S_N	D_0
1	$\overline{S_1}$	S_1	D_1
2	S_P	$\overline{S_P}$	D_2

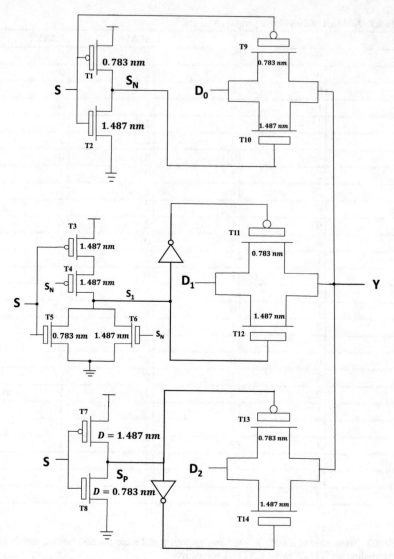

Fig. 4.1 CNTFET-based 3×1 multiplexer

Table 4.4 Truth table of basic unary operators

A	A_P	A_N	$\overline{A}$	A^1	A^2	A_0	A_1	A_2
0	2	2	2	1	2	2	0	0
1	2	0	1	2	0	0	2	0
2	0	0	0	0	1	0	0	2

SUM output of a ternary half-adder can be obtained from Table 4.1. We therefore have Eq. (4.1) omitting the terms involving AND (min) with 0.

$$SUM = 1 \cdot B_0 A_1 + 2 \cdot A_2 B_0 + 1 \cdot B_1 A_0$$
$$+2 \cdot A_1 B_1 + 2 \cdot A_0 B_2 + 1 \cdot A_2 B_2 \tag{4.1}$$

Remark 4.2 The equation for sum can be verified as follows. We limit to a few inputs although the equations hold for various combinations.

Consider the fifth row in Table 4.1 corresponding to $A = B = 1$. This implies $A_0 = A_2 = B_0 = B_2 = 0$ and $A_1 = B_1 = 2$. Substituting into the right-hand side of Eq. (4.1), we have $B_0 A_1$, $A_2 B_0$, $B_1 A_0$, $A_0 B_2$ and $A_2 B_2$ evaluating to 0. Hence, the terms $min(1, B_0 A_1)$, $min(2, A_2 B_0)$, $min(1, B_1 A_0)$, $min(2, A_0 B_2)$ and $min(1, A_2 B_2)$ evaluate to 0 while $2 \cdot A_1 B_1$ evaluates to $min(2, 2) = 2$ since $A_1 = B_1 = 2$ when $A = B = 1$ (we note that A_1 is a decisive operator as shown in Table 4.4; same holds for B_1).

Now consider the ninth row in Table 4.1 corresponding to $A = B = 2$. This implies $A_0 = A_1 = B_0 = B_1 = 0$ and $A_2 = B_2 = 2$. Substituting into the right-hand side of Eq. (4.1), we have $B_0 A_1$, $A_2 B_0$, $B_1 A_0$, $A_0 B_2$ and $A_1 B_1$ evaluating to 0. Hence, the terms $min(1, B_0 A_1)$, $min(2, A_2 B_0)$, $min(1, B_1 A_0)$, $min(2, A_0 B_2)$ and $min(2, A_1 B_1)$ evaluate to 0 while $1 \cdot A_2 B_2$ evaluates to $min(1, 2) = 1$.

Analogous to the derivation for sum, we have the expression for carry for a ternary half-adder given by Eq. (4.2).

$$CARRY = 1 \cdot A_2 B_1 + 1 \cdot A_1 B_2 + 1 \cdot A_2 B_2 \tag{4.2}$$

Similarly, the two outputs of a ternary full-adder (namely SUM and CARRY) can be obtained directly from Table 4.2. The corresponding equations are (4.3) and (4.4).

$$SUM = [A_0 B_0 C_1 + A_0 B_1 C_0 + A_0 B_2 C_2 + A_1 B_0 C_0 + A_1 B_1 C_2$$
$$+A_1 B_2 C_1 + A_2 B_0 C_2 + A_2 B_1 C_1 + A_2 B_2 C_0] \cdot 1$$
$$+[A_0 B_0 C_2 + A_0 B_1 C_1 + A_0 B_2 C_0 + A_1 B_0 C_1 + A_1 B_1 C_0$$
$$+A_1 B_2 C_2 + A_2 B_0 C_0 + A_2 B_1 C_2 + A_2 B_2 C_1] \cdot 2 \tag{4.3}$$

$$CARRY = [A_0 B_1 C_2 + A_0 B_2 C_1 + A_0 B_2 C_2 + A_1 B_0 C_2 + A_1 B_1 C_1$$
$$+A_1 B_1 C_2 + A_1 B_2 C_0 + A_1 B_2 C_1 + A_1 B_2 C_2 + A_2 B_0 C_1$$
$$+A_2 B_0 C_2 + A_2 B_1 C_0 + A_2 B_1 C_1 + A_2 B_1 C_2 + A_2 B_2 C_0$$
$$+A_2 B_2 C_1] \cdot 1 + 2 \cdot A_2 B_2 C_2 \tag{4.4}$$

We now proceed to discuss efficient design of a ternary half-adder and a ternary full-adder.

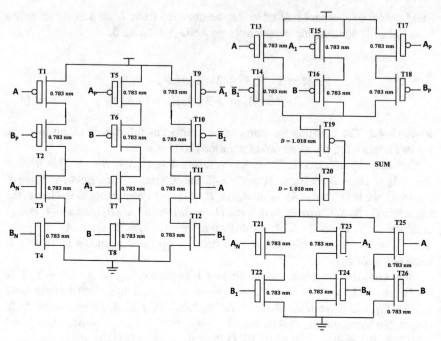

Fig. 4.2 SUM (Output) of ternary half-adder (without multiplexers)

4.4 Proposed CNTFET-Based Ternary Half-Adder

We first present a design for a ternary half-adder that does not employ multiplexers. The ternary half adder SUM and CARRY are computed using the circuits shown in Figs. 4.2 and 4.3 respectively. The construction and operation of the half-adder circuits are as follows. We begin with the one for SUM shown in Fig. 4.2. The circuit is constructed with CNTFETs of two diameters, namely 0.783 and 1.018 nm. An n-type CNTFET with 0.783 nm diameter will be ON for logic '2' while it is OFF for inputs corresponding to logic '0' and '1'. Similarly, the p-type CNTFET with a diameter of 0.783 nm will be ON for input voltage corresponding to logic '0' while it is OFF for inputs corresponding to logic '1' and '2'. The SUM output of the ternary half adder has three possibilities, namely '0', '1' and '2'. SUM will be '0' for inputs A = '0' B = '0', A = '1' B = '2' and A = '2' B = '1'. Now, the n-type CNTFETs $T3$ and $T4$ correspond to the input pair A = '0' B = '0'. Hence, these two CNTFETs $T3$ and $T4$ are connected to A_N and B_N, so these two CNTFETs will be ON for A = 0 and B = 0. This results in the output SUM being connected to ground, hence 0V (i.e., logic '0'). Similarly, the pairs $T7$, $T8$ and $T11$, $T12$ correspond to the input pair A = '1' B = '2' and A = '2' B = '1' respectively. These CNTFETs ($T7$ and $T8$) are connected to A_1 and B respectively. This pair will be ON for the combination of

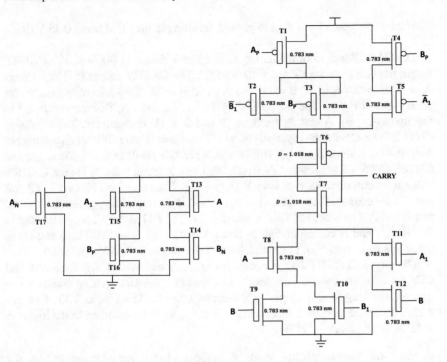

Fig. 4.3 CARRY (Output) of ternary half-adder (without multiplexers)

inputs given by A = '1' and B ='2'. As a result, the output SUM is connected to ground, hence we obtain 0 V (logic '0').

Suppose we have inputs A = '0' B = '2', A = '1' B = '1' and A = '2' B = '0'. The output SUM is '2'. In this case, the p-type CNTFETs $T1$, $T2$ whose inputs are connected to A and B_P correspond to the input pair A = '0' B = '2'. This pair of CNTFETs is ON only for the input A = '0' and B = '2', causing SUM to be connected to V_{DD} (logic '2'). Similarly, the CNTFETs $T5$, $T6$ and $T9$, $T10$ are connected to inputs A_P, B and $\overline{A_1}$, $\overline{B_1}$ respectively for inputs A = '2' B = '0' and A = '1' B='1'.

Now consider the cases of SUM being '1' for inputs A = '0' B = '1', A = '1' B = '0' and A = '2' B = '2'. The p-type CNTFETs $T13$, $T14$ and n-type CNTFETs $T21$, $T22$ are connected to A, $\overline{B_1}$, A_N and B_1 respectively corresponding to the inputs A='0' and B='1'. These CNTFETs $T13$, $T14$, $T21$ and $T22$ are ON for inputs A = '0' and B = '1'. Since these CNTFETs are ON, it results in the diode connected CNTFETs $T19$, $T20$ to become ON. Hence, this results in a path from V_{DD} to ground, leading to the output (for SUM) of 0.45 V (logic '1'). Similarly, the p-type CNTFETs $T15$, $T16$ and n-type CNTFETs $T23$, $T24$ are connected to $\overline{A_1}$, B, A_1 and B_N respectively corresponding to the inputs A = '1' and B = '0'. The other CNTFETs, namely $T17$, $T18$, $T25$ and $T26$ are connected to A_P, B_P, A and B respectively corresponding to inputs A = '2' and B = '2'. These two cases lead (the circuit) to a path from V_{DD} to ground, resulting in the SUM being 0.45 V (logic '1').

The CARRY output of the ternary half adder is shown in Fig. 4.3. The CARRY output has two possible values of '0' and '1'. The CARRY output is '1' for inputs A = '1' B = '2', A = '2' B = '1' and A = '2' B = '2'. CARRY (output) is '0' for other cases of inputs. The p-type CNTFETs $T1$, $T2$ and $T8$, $T10$ are connected to the inputs A_P, $\overline{B_1}$, A and B_1 for A = '2' and B = '1' respectively. These pairs of CNTFETs are ON for the input pair of A = '2' and B = '1' and OFF for all other input combinations. Since these CNTFETs which are ON result in the diode connected CNTFETs $T6$, $T7$ to become ON, a path from V_{DD} to ground results. Hence, CARRY will get an output voltage of 0.45 V (logic '1'). The other CNTFETs $T1$, $T3$, $T8$ and $T9$ are connected to inputs A_P, B_P, A and B corresponding to inputs A = '2' and B = '2'. The other CNTFETs, namely $T4$, $T5$, $T11$ and $T12$ are connected to B_P, $\overline{A_1}$, A_1 and B corresponding to input A = '1' and B = '2'. These two cases result in a path from V_{DD} to ground, leading the output to 0.45 V (logic '1').

The n-type CNTFET $T17$ is connected to A_N, which will be ON for A='0' and OFF for all other inputs. This results in the output CARRY getting connected to ground, resulting in 0 V (logic '0'). Similarly, the CNTFET pairs $T13$, $T14$ and $T15$, $T16$ are connected to inputs A, B_N and A_1, B_P corresponding to the inputs A = '2' B = '0' and A = '1' B = '0'.

Remark 4.3 The complexity details of the ternary half adder presented in Figs. 4.2 and 4.3 are as follows. SUM requires 26 CNTFETs while CARRY requires 17 CNT-FETs. The control signals, namely A_P, B_P, A_N, B_N, A_1, B_1, $\overline{A_1}$ and $\overline{B_1}$, require 2, 2, 2, 2, 4, 4, 2 and 2 CNTFETs respectively. Hence, the ternary half adder requires a total of 63 CNTFETs.

We now consider a multiplexer-based design. A 3×1 ternary multiplexer-based design for a ternary half-adder that takes certain unary operators as inputs can be developed. It is a modification of the design in [3]. We begin by rewriting the basic equations for SUM and CARRY (presented earlier) in a form that is appropriate for multiplexer-based realization. Consider Eqs. (4.1) and (4.2). We can rewrite them such that B is used as the 'select' signal for a multiplexer as given by Eqs. (4.5) and (4.6) respectively.

$$SUM = B_0[1 \cdot A_1 + 2 \cdot A_2] + B_1[1 \cdot A_0 + 2 \cdot A_1]$$
$$+ B_2[2 \cdot A_0 + 1 \cdot A_2] \tag{4.5}$$
$$CARRY = B_1[1 \cdot A_2] + B_2 \cdot [1 \cdot A_1 + 1 \cdot A_2] \tag{4.6}$$

A direct realization of merely "$1 \cdot A_0 + 2 \cdot A_1$" in Eq. (4.5) requires 48 CNTFETs (16 CNTFETs for each AND and 16 for OR). However, taking advantage of relationships between unary operators, the transistor count can be substantially reduced. In particular, we note that $1 \cdot A_1 + 1 \cdot A_2$ is the same as $1 \cdot \overline{A^2}$. Similarly, one can

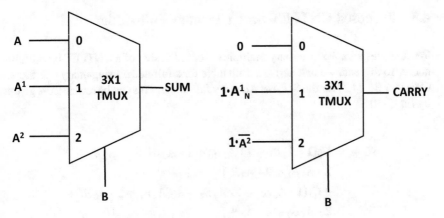

Fig. 4.4 Ternary MUX-based half adder using unary operators

obtain simplified forms for $[2 \cdot A_0 + 1 \cdot A_2]$, $[1 \cdot A_0 + 2 \cdot A_1]$ and so on. This leads to Eqs. (4.7) and (4.8). The multiplexer-based design is shown in Fig. 4.4.

$$SUM = B_0 \cdot A + B_1 \cdot A^1 + B_2 \cdot A^2 \qquad (4.7)$$
$$CARRY = B_0 \cdot 0 + B_1 \cdot [1 \cdot A_N^1] + B_2 \cdot [1 \cdot \overline{A^2}] \qquad (4.8)$$

Remark 4.4 In the expression for CARRY, we have used A_N^1 where the suffix N corresponds to negative ternary inversion applied to A^1. Note that A^1 is given by $(1, 2, 0)$ while A_N^1 corresponds to $(0,0,2)$.

In place of $1 \cdot \overline{A^2}$, one can also have $1 \cdot \overline{A}$ in the expression for CARRY for a half-adder. Other choices are also possible and it is worth noting that the CNTFET-count changes substantially depending on the choice of inputs.

Remark 4.5 The complexity details of the multiplexer based ternary half adder are as follows. From the circuits already presented for unary operators, it is clear that the cycle operators, A^1 and A^2, require three CNTFETs each. The other two operators, namely $1 \cdot A_N^1$ and $1 \cdot \overline{A^2}$, require 2 CNTFETs each. In addition, these circuits require one PTI and NTI, each of which requires 2 CNTFETs (see also Fig. 4.1 where from a select signal S, we derive S_N and S_P). The input to the multiplexers requires a total of 14 CNTFETs. The ternary multiplexer requires 6 CNTFETs for transmission gate and 12 CNTFETs for the selection of the transmission gate. The ternary half adder has two 3×1 multiplexers and both the multiplexers have the same control signal (derived from input 'B'). The control signal requires 12 CNTFETs while 6 CNTFETs per multiplexer are required for data passing (transmission gates) hence we need 24 CNTFETs altogether for this component of the circuit. Overall therefore, the multiplexer-based ternary half adder requires a total of 38 CNTFETs.

4.5 Proposed CNTFET-Based Ternary Full-Adder

We first present a 3×1 ternary multiplexer-based design of a CNTFET-based full-adder. To this end, we first derive a truth table for a full-adder using unary operators. Consider Eq. (4.3) for the SUM of a ternary full-adder. This can be rewritten as given by Eq. (4.9).

$$
\begin{aligned}
SUM = {} & C_0[1 \cdot A_0 B_1 + 1 \cdot A_1 B_0 + 1 \cdot A_2 B_2 + 2 \cdot A_0 B_2 + \\
& 2 \cdot A_1 B_1 + 2 \cdot A_2 B_0] \\
& + C_1[1 \cdot A_0 B_0 + 1 \cdot A_1 B_2 + 1 \cdot A_2 B_1 + 2 \cdot A_0 B_1 + \\
& 2 \cdot A_1 B_0 + 2 \cdot A_2 B_2] \\
& + C_2[1 \cdot A_0 B_2 + 1 \cdot A_1 B_1 + 1 \cdot A_2 B_0 + 2. \cdot A_0 B_0 + \\
& 2. \cdot A_1 B_2 + 2. \cdot A_2 B_1]
\end{aligned} \tag{4.9}
$$

Equation (4.9) can be rewritten as given by Eq. (4.10).

$$
\begin{aligned}
SUM = {} & C_0[B_0(1 \cdot A_1 + 2 \cdot A_2) + B_1(1 \cdot A_0 + 2 \cdot A_1) \\
& + B_2(1 \cdot A_2 + 2 \cdot A_0)] \\
& + C_1[B_0(1 \cdot A_0 + 2 \cdot A_1) + B_1(1 \cdot A_2 + 2 \cdot A_0) \\
& + B_2(1 \cdot A_1 + 2 \cdot A_2)] \\
& + C_2[B_0(1 \cdot A_2 + 2 \cdot A_0) + B_1(1 \cdot A_1 + 2 \cdot A_2) \\
& + B_2(1 \cdot A_0 + 2 \cdot A_1)]
\end{aligned} \tag{4.10}
$$

Now noting that we can replace (i) $(1 \cdot A_1 + 2 \cdot A_2)$ by A (ii) $(1 \cdot A_0 + 2 \cdot A_1)$ by A^1 and (iii) $(1 \cdot A_2 + 2 \cdot A_0)$ by A^2, we have Eq. (4.11).

$$
\begin{aligned}
SUM = {} & C_0[B_0 \cdot A + B_1 \cdot A^1 + B_2 \cdot A^2] + C_1[B_0 \cdot A^1 + B_1 \cdot A^2 + B_2 \cdot A] \\
& + C_2[B_0 \cdot A^2 + B_1 \cdot A + B_2 \cdot A^1]
\end{aligned} \tag{4.11}
$$

Similarly, the expression for CARRY for a ternary full-adder given by Eq. (4.4) can be rewritten in the form given by Eq. (4.12).

$$
\begin{aligned}
CARRY = {} & C_0[B_0 \cdot 0 + B_1 \cdot (1 \cdot A_N^1) + B_2 \cdot (1 \cdot \overline{A^2})] \\
& + C_1[B_0 \cdot (1 \cdot A_N^1) + B_1 \cdot (1 \cdot \overline{A^2}) + B_2 \cdot 1] \\
& + C_2[B_0 \cdot (1 \cdot \overline{A^2}) + B_1 \cdot 1 + B_2(1 + A_N^1)]
\end{aligned} \tag{4.12}
$$

Table 4.5 Ternary full adder truth table using the unary operators

C	B	A	CARRY	SUM
0	0	A	0	A
0	1	A	$1 \cdot A_N^1$	A^1
0	2	A	$1 \cdot \overline{A^2}$	A^2
1	0	A	$1 \cdot A_N^1$	A^1
1	1	A	$1 \cdot \overline{A^2}$	A^2
1	2	A	1	A
2	0	A	$1 \cdot \overline{A^2}$	A^2
2	1	A	1	A
2	2	A	$1 + A_N^1$	A^1

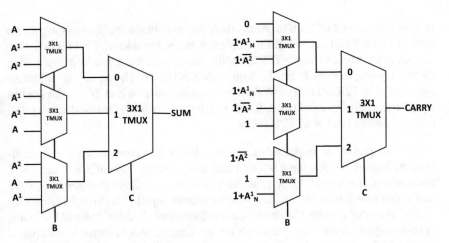

Fig. 4.5 Ternary MUX-based full adder using unary operators

It can be observed that Eqs. (4.11) and (4.12) lead to a two-level multiplexer-based realization. Table 4.5 captures the behaviour of a ternary full-adder in terms of unary operators.

Remark 4.6 The design based on two "levels" of multiplexers for a ternary full-adder can be thought of as an extension of the single-level solution for a half-adder. In the former, there are three inputs to be handled (as opposed to two in a half-adder), hence it is natural to let two of the three inputs become select signals. Appropriate choice of unary operators further enables obtaining a low-complexity design.

Remark 4.7 The complexity details of the 3×1 multiplexer-based ternary fulladder (given by Fig. 4.5) are as follows. The cycle operators, A^1 and A^2, require three CNTFETs each. The other operators, namely $1 \cdot A_N^1$, $1 \cdot \overline{A^2}$ and $1 + A_N^1$ take 2,2 and 5 CNTFETs respectively. In addition, these circuits require one PTI and NTI, each

Table 4.6 Truth table for deriving a new control signal X, which is the sum of two inputs

A	B	X
0	0	0
0	1	1
0	2	2
1	0	1
1	1	2
1	2	0
2	0	2
2	1	0
2	2	1

of which requires 2 CNTFETs. Altogether, the input to the multiplexers requires a total of 19 CNTFETs. The eight multiplexers altogether take 48 CNTFETs for the transmission gates and 12 CNTFETs for the selection of the transmission gate. Two control signals, namely B and C, require 24 CNTFETs. The ternary multiplexers therefore take 72 CNTFETs (48 for the transmission gate and 24 for the control signal from two inputs). Hence, the 3 × 1 ternary multiplexer-based ternary full adder requires a total of 91 CNTFETs.

We now present an improved design that does not require several 3 × 1 multiplexers. Instead, we develop a solution that uses a combination of 3 × 1 and 2 × 1 multiplexers. The key idea is to generate a 'new' control signal (denoted by X). We begin with a truth table that depicts how the control signal X is derived (Table 4.6).

We know that a ternary full adder can be derived by choosing some of the inputs as select signals to the ternary multiplexer. An alternative is to derive a new control signal which acts as a select line to derive the (overall) SUM and CARRY. Let the sum of two inputs $\{A, B\}$ be 'X'. The three possible values are 0, 1 and 2. Hence, the sum of three ternary numbers can be calculated as per Eq. (4.13).

$$SUM = \begin{cases} C, & \text{if } X = 0 \\ C^1, & \text{if } X = 1 \\ C^2, & \text{if } X = 2 \end{cases} \tag{4.13}$$

Similarly, CARRY can be computed from the signal 'X'. While computing CARRY, we need to consider the second input 'B' as well.

$$CARRY = \begin{cases} 0, & \text{if } B = 0 \text{ and } X = 0 \\ 1, & \text{if } B = 1, 2 \text{ and } X = 0 \\ 1 \cdot \overline{C_P}, & \text{if } B = 0, 1 \text{ and } X = 1 \\ 1 + \overline{C_P}, & \text{if } B = 2 \text{ and } X = 1 \\ 1 \cdot \overline{C_N}, & \text{if } X = 2 \end{cases} \qquad (4.14)$$

The ternary full adder can then be implemented using these equations. We note that SUM expressed in Eq. (4.13) has three values for three different values of 'X'. Therefore, SUM can be realized using a 3×1 multiplexer. Note that CARRY has an extra condition (with the second input 'B'). We can list the various possibilities as follows.

- When $X = 0$, CARRY can be '0' and '1' for $B = 0$ and $B = 1, 2$ respectively. Here, a 2×1 multiplexer can be used with 'B' as the select signal.
- When $X = 1$, two different carry outputs are possible for $B = 0, 1$ and $B = 2$. This can be implemented using a 2×1 multiplexer.
- Finally, a 3×1 multiplexer with 'X' as the control signal will give us the CARRY output of the ternary full adder.

This results in an improved full adder design using just three 3×1 multiplexers and two 2×1 multiplexers as shown in Fig. 4.6. This design leads to reduction of (approximately) 15% CNTFETs (from the one that uses only 3×1 multiplexers). This is established via Remark 4.8.

Remark 4.8 . The complexity for the improved design in Fig. 4.6 can be analysed as follows. The total CNTFET count has two components: (i) CNTFETs required for realization of the unary operators and (ii) the CNTFET count for the multiplexers themselves. The cycle operators, A^1 and A^2, require three CNTFETs each. These operators require PTI and NTI (as A_P and A_N), which take 2 CNTFETs each. The remaining operators, namely $1 \cdot A_N^1$, $1 \cdot \overline{A^2}$ and $1 + \overline{A_N^1}$ take 2, 2 and 5 CNTFETs respectively. In addition, these circuits require one PTI and NTI, each of which requires 2 CNTFETs. Altogether, the input to the multiplexers requires a total of 27 CNTFETs.

Since there are altogether three 3×1 multiplexers, they need 18 CNTFETs for the transmission gates. The control signals require 12 CNTFETs for the selection of the transmission gate. Two controls namely, B and X, as shown in Fig. 4.6, take 24 CNTFETs. The two 2×1 ternary multiplexers take 8 CNTFETs for the transmission gates. Hence, the overall transistor count for this improved design is 77. This corresponds to a reduction of 14 CNTFETs from the one that was based entirely on 3×1 multiplexers.

Remark 4.9 In order to see the benefits of the unary operator-based approach (with respect to transistor count), we present some comparisons with prior work. The decoder-based ternary half adder presented in [2] consumes 116 CNTFETs for the sum and carry outputs.

With respect to prior full-adder designs, the one in [2] requires 318 CNTFETs. A recent design presented in [4] requires 144 CNTFETs.

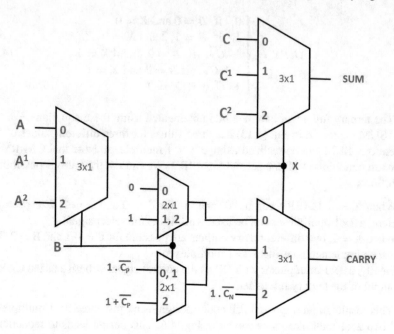

Fig. 4.6 Ternary full adder with a combination of 2×1 and 3×1 multiplexers

Remark 4.10 Another aspect that is of interest is low-power design (with slightly higher transistor count). For instance, power reduction can be accomplished via diode-connected CNTFETs in unary circuits while realizing logic 1.

4.6 Summary

In this chapter, we have pursued efficient design of single trit adders. The designs are based on judicious selection of unary operators and careful use of multiplexers. In the next chapter, we extend the study to a multi-trit adder.

References

1. Dhande, A.P., Ingole, V.T.: Design and Implementation of 2-bit ternary ALU slice. Proceedings of International Conference IEEE-Science Electronic Technology Information Telecommunication, pp. 17–21 (2005)
2. Lin, S., Kim, Y.B., Lombardi, F.: CNTFET-based design of ternary logic gates and arithmetic circuits. IEEE Trans. Nanotechnol. **10**(2), 217–225 (2011)
3. Srinivasu, B., Sridharan, K.: Carbon nanotube FET-based low-delay and low-power multidigit adder designs. IET-Circuits, Devices Syst. **11**(4), 352–364 (2017)
4. Keshavarzian, P., Sarikhani, R.: A novel CNTFET-based ternary full adder. Circuits, Syst. Signal Process. **33**, 665–679 (2014)

Chapter 5
CNTFET-Based Design of a Multi-ternary Digit Adder

In the previous chapter, we have explored efficient design of CNTFET-based circuits for single ternary digit addition. In this chapter, we consider an extension. In particular, we study multi-ternary digit addition. We develop a multi-ternary digit adder based on the notion of conditional sum [1]. The problem addressed in this chapter is also discussed in [2].

The design presented takes advantage of single-trit half and full-adders presented in the previous chapter. The circuit also involves modified half-adders (that add two "variable-digits" and a constant).

5.1 Overview of a CNFTET-Based Ternary Conditional Sum Adder

We recall that, in the binary setting, there are various multibit adders. These include the ripple carry adder, carry lookahead adder, conditional sum adder and prefix adders. Of these, the ripple carry adder (RCA) turns out to be the simplest and involves a cascade of several full adders. In an RCA, carry propagation takes place from one full adder to the next leading to substantial delay in production of the 'final' carry and sum outputs. Modifications to reduce the delay have been proposed and these include the carry lookahead adder, conditional sum adder and so on. In a conditional sum adder (CSA), all possible outputs are calculated in advance by assuming the input carry to be 0 or 1. An extension of this idea to the ternary case corresponds to assumption of the input carry to be '0', '1' or '2'. The actual output is selected using multiplexers.

In a ternary conditional sum adder, for a given input A and B, we calculate all the possible sum and carry values in the first stage by considering the three possibilities for the input carry, i.e., $C_{in} = 0$, $C_{in} = 1$ and $C_{in} = 2$. Hence, in the first stage of the

© Springer Nature Switzerland AG 2020
K. Sridharan et al., *Low-Complexity Arithmetic Circuit Design in Carbon Nanotube Field Effect Transistor Technology*, Carbon Nanostructures,
https://doi.org/10.1007/978-3-030-50699-5_5

CSA, three types of adders are required which will perform the addition operations namely $A_i + B_i$, $A_i + B_i + 1$ and $A_i + B_i + 2$. Here, A_i and B_i correspond to the ith digit of the given inputs A and B respectively.

One approach to handle $A_i + B_i$, $A_i + B_i + 1$ and $A_i + B_i + 2$ is by a half adder for $A_i + B_i$ and two full adders, one each for $A_i + B_i + 1$ and $A_i + B_i + 2$. On the other hand, noting that the input carry to the adders is fixed as '0', '1' or '2', we can have a more efficient solution via a half adder and modified half-adders. A modified half adder is a two-digit adder which corresponds to a variation of the classical half-adder. One modified half-adder computes $A_i + B_i + 1$ while the other computes $A_i + B_i + 2$. We will denote the former by MHA-1 and the latter by MHA-2. We next discuss the role of unary operators in multi-trit addition.

5.2 Key Unary Operators for Getting a Low-Complexity Multi-trit Conditional Sum Adder

As for single digit adders, careful selection of unary operators plays an important role in obtaining low complexity designs. For multi-trit addition, the key unary operators are given in Table 5.1.

5.3 Efficient Realization of the Components of a Ternary Conditional Sum Adder

As indicated earlier, two variations of the classical half-adder play a valuable role in obtaining a low-complexity multi-trit adder. These correspond to MHA-1 and MHA-2. We present the details of these two modified half-adders next.

5.3.1 Realizing the Ternary Modified Half Adder MHA-1

The truth table for MHA-1 is given in Table 5.2. This can be expressed using unary operators and the corresponding truth table for MHA-1 is shown in Table 5.3. A multiplexer-based design of MHA-1 is shown in Fig. 5.1a.

Table 5.1 Key unary operators for multi-trit adder design

A	A^1	A^2	A_N^1	$1 \cdot A_N^1$	$\overline{A^2}$	$1 \cdot \overline{A^2}$	$1 + A_2$
0	1	2	0	0	0	0	1
1	2	0	0	0	2	1	1
2	0	1	2	1	1	1	2

Table 5.2 Truth table for MHA-1; sum and carry correspond to the outputs of $A + B + 1$

A	B	Carry	Sum
0	0	0	1
0	1	0	2
0	2	1	0
1	0	0	2
1	1	1	0
1	2	1	1
2	0	1	0
2	1	1	1
2	2	1	2

Table 5.3 Truth table for MHA-1 in terms of unary operators

A	B	Carry	Sum
A	0	$1 \cdot A_N^1$	A^1
A	1	$1 \cdot \overline{A^2}$	A^2
A	2	1	A

5.3.2 Realizing Ternary Modified Half Adder MHA-2

The modified half adder MHA-2 is given by Tables 5.4 and 5.5. A multiplexer-based design for MHA-2 is shown in Fig. 5.1b.

5.4 Structure of a 3-Trit Conditional Sum Adder

In order to see how the single-trit adders fit into the overall CSA structure, we depict a 3-trit conditional sum adder in Fig. 5.2. We note that a dual 6×2 multiplexer can be realized using 3×1 multiplexers. Scaling the design to larger size ternary conditional sum adders is via usage of the 3-trit CSA blocks.

An example showing the operation of the 3-trit conditional sum adder is given in Fig. 5.3. It may be noted that the addition of 021_3 (i.e., $1 \times 3^0 + 2 \times 3^1$) and 022_3 (i.e., $2 \times 3^0 + 2 \times 3^1$) is performed in multiple steps in the conditional sum addition circuitry to obtain the sum, namely 15_{10}.

Fig. 5.1 Multiplexer-based
realization of modified
half-adders

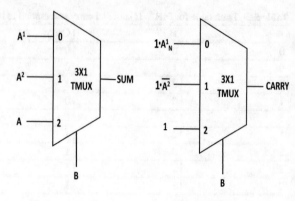

(a) MUX-based MHA-1 $(A + B + 1)$

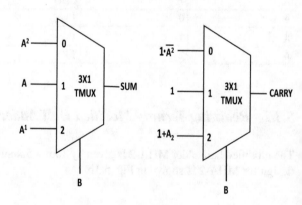

(b) MUX-based MHA-2 $(A + B + 2)$

Table 5.4 Truth table for MHA-2; sum and carry correspond to the outputs of $A + B + 2$

A	B	Carry	Sum
0	0	0	2
0	1	1	0
0	2	1	1
1	0	1	0
1	1	1	1
1	2	1	2
2	0	1	1
2	1	1	2
2	2	2	0

Table 5.5 Truth table for MHA-2 in terms of unary operators

A	B	Carry	Sum
A	0	$1 \cdot \overline{A}^2$	A^1
A	1	1	A^2
A	2	$1 + A_2$	A

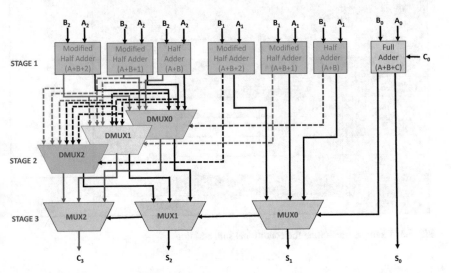

Fig. 5.2 Structure of 3-trit conditional sum adder realized using MUXes and one-trit adders

5.5   Determining the CNTFET-Count for the 3-Trit Conditional Sum Adder

From the structure of the 3-trit conditional sum adder, it is evident that the design requires full and half adders, custom modified half adders and a few multiplexers. From single trit adder design discussions, we know that a ternary full adder requires 91 CNTFETs while a half adder requires 38 CNTFETs (24 CNTFETs for the multiplexers and 14 CNTFETs for the unary operators). One can observe from the designs of modified half adders MHA-1 and MHA-2 that these require many unary operators which are also present in the half adder design. However, in MHA-2 design, we have a unary operator, $1 + A_2$. Since the input operators involved in the half adder and modified adders are same, we can share the unary operator among these adders for the digits A_1, B_1 and A_2, B_2. Moreover, the multiplexer control signals are also same for these adders. Also, MHA-1 requires 12 CNTFETs for the transmission gate in the two multiplexers while MHA-2 requires 12 CNTFETs for the transmission gate in the multiplexer and 5 CNTFETs for the unary operator $(1 + A_2)$. Altogether the half adder, MHA-1 and MHA-2 require 67 CNTFETs (38 for half adder, 12 for

0	2	1		A	7
0	2	2		B	8
	0	1	0	S^0	1st
	0	1	1	C^0	S
1	2	-		S^1	T
0	1	-		C^1	A
2	0	-		S^2	G
0	2	-		C^2	E
1	1	0		DMUX0	2nd
0	-	1			S
1	2	-		DMUX1	T
0	-	-			A
2	0	-			G
0	-	-		DMUX2	E
1	2	0		SUM	15
0				CARRY	

Fig. 5.3 Example illustrating the conditional sum addition

MHA-1 and 17 for MHA-2). With regard to the DMUX (dual multiplexer), we note that there is one control signal and there are two outputs. The control signal takes 12 CNTFETs while the transmission gates for the two outputs require 12 CNTFETs. Hence, a DMUX takes 24 CNTFETs. The final multiplexers (MUX0, MUX1 and MUX2) have same control signal and three outputs. These multiplexers requires 30 CNTFETs.

Therefore, the total CNTFET count is 327 and this can be expressed in terms of the components as follows.

- Full adder (for $A_0 + B_0 + C_0$) requires 91 CNTFETs.
- The adders (for $A_1 + B_1$, $A_1 + B_1 + 1$ and $A_1 + B_1 + 2$) require 67 CNTFETs. Two such adders are present, hence we have altogether 134 CNTFETs. The first stage therefore requires 225 (134 + 91) CNTFETs.
- The dual multiplexers in the second stage require 72 CNTFETs.
- The multiplexers in the third stage require 30 CNTFETs.

5.6 Summary

We have presented the design of a ternary conditional sum adder and discussed its CNTFET-based realization in this chapter. The design takes advantage of low-complexity single trit adder solutions presented earlier.

While careful selection of unary operators plays an important role in obtaining low-complexity designs, identifying an optimal combination of unary operators is open and can be pursued to further reduce the CNTFET count.

References

1. Koren, I.: Computer Arithmetic Algorithms. A.K. Peters Ltd. (2002)
2. Srinivasu, B., Sridharan, K.: Carbon nanotube FET-based low-delay and low-power multidigit adder designs. IET-Circuits, Devices Syst. **11**(4), 352–364 (2017)

Chapter 6
CNTFET-Based Design of a Ternary Multiplier

We have discussed the design of CNTFET-based adders in the earlier chapters. In this chapter, we present the design of a ternary multiplier. We note that the design of a single-digit multiplier itself is non-trivial in the ternary setting. A somewhat related study has been pursued in [1].

6.1 Multiplying Two Ternary Digits

In the binary case, the product of a pair of bits is obtained simply by an AND operation. However, in the ternary case, the circuitry for realizing the product of a pair of (ternary) digits is much more complex. The reason is that the product of two ternary digits is not necessarily a single ternary digit. Table 6.1 shows the various possibilities that arise when multiplying a ternary digit A by another ternary digit B.

Remark 6.1 A typical entry in Table 6.1 can be verified as follows: Let A and B be given by 2 and 2 respectively. The product of these gives 4 (in base-10) which can be expressed as $1 \times 3^1 + 1 \times 3^0$ hence, the product (trit) is 1 while the carry (trit) is also 1.

6.2 Single Ternary Digit Multiplier Design

As observed in Sect. 6.1, given a pair of ternary digits as input, we produce two quantities, namely a *product* and a *carry*. We begin with a direct scheme for this purpose and then move on to a low-complexity ternary multiplexer-based design.

© Springer Nature Switzerland AG 2020
K. Sridharan et al., *Low-Complexity Arithmetic Circuit Design in Carbon Nanotube Field Effect Transistor Technology*, Carbon Nanostructures,
https://doi.org/10.1007/978-3-030-50699-5_6

Table 6.1 Table for Single Ternary Digit Multiplication

B	A	Product	Carry
0	0	0	0
0	1	0	0
0	2	0	0
1	0	0	0
1	1	1	0
1	2	2	0
2	0	0	0
2	1	2	0
2	2	1	1

From Table 6.1, we can obtain Eqs. (6.1) and (6.2) for the *product* and *carry* respectively. As before, we note that when $B = 1$, the value of B_2 is 2.

$$product = 1 \cdot B_1 A_1 + 2 \cdot B_1 A_2 + 2 \cdot B_2 A_1 + 1 \cdot B_2 A_2 \qquad (6.1)$$

$$carry = 1 \cdot B_2 A_2 \qquad (6.2)$$

A direct realization of the *product* alone in Eq. (6.1) requires substantial number of CNTFETs and the details are as follows. Note that the product has two possible values, namely logic '1' or logic '2'. It is realized using the circuit shown in Fig. 6.1. The circuit operates as follows.

- The two p-type CNTFETs, namely $T1$ and $T2$, have their gate inputs connected to A_P and $\overline{B_1}$ respectively. These CNTFETs have a diameter of 0.783 nm and will be ON only for an input corresponding to logic '0' (and OFF for other inputs). This pair of CNTFETs will be ON for input pair A = '2' and B = '1'. As a result, the output *product* connected to V_{DD}, will be at logic '2'. Similarly, the other p-type CNTFETs, namely, $T2$ and $T4$, have gate controlled by $\overline{A_1}$ and B_P. This pair will be ON for input A = '1' and B = '2', leading the output to V_{DD} (i.e., logic '2').
- Consider the p-type CNTFETs $T5$ and $T6$ which are controlled by $\overline{A_1}$ and $\overline{B_1}$ with diameter 0.783 nm. This pair will be ON for input A = '1' and B = '1'. The n-type CNTFETs $T13$ and $T14$ which are controlled by A_1 and B_1, have diameter of 0.783 nm. This pair ($T13$, $T14$) will be ON for A = '1' and B = '1'. Since the CNTFETs $T5$, $T6$, $T13$ and $T14$ are ON for A = '1' and B = '1', this leads to the two diode connected CNTFETs $T9$ and $T10$ becoming ON. As a result, the circuit behaves like a voltage divider. This results in the output of $\frac{V_{DD}}{2}$ (i.e., logic '1'). Similar arguments hold for the CNTFETs $T8$, $T9$, $T15$ and $T16$ for inputs A = '2' and B = '2'.
- The two n-type CNTFETs namely, $T11$ and $T12$, are in parallel and are controlled by A_N and B_N, with diameters of 0.783 nm. CNTFET $T11$ will be ON when A =

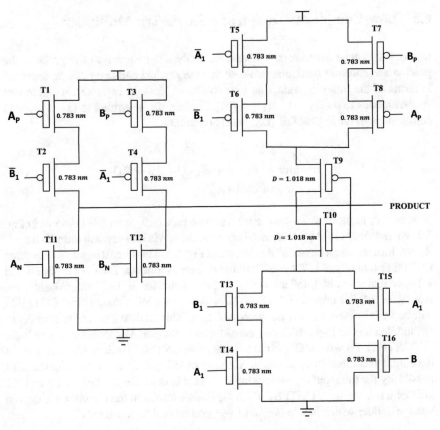

Fig. 6.1 CNTFET-based direct realization of the product of the single trit multiplier

'0', leading the output *product* to connect to ground (i.e., logic '0'). The CNTFET $T12$ will be ON for B = '0', leading the output to '0'.

The CNTFET-based direct realization of the *product* of the single trit multiplier requires 36 CNTFETs. This includes the 16 CNTFETs in the circuit shown plus the ones for the control signals such as A_P, B_P, A_N, B_N, A_1, B_1, $\overline{A_1}$ and $\overline{B_P}$. In particular, the control signals require 20 CNTFETs. The second output of a single trit multiplier, namely *carry*, requires few more CNTFETs. Hence, the direct solution requires substantial number of CNTFETs so it is of interest to seek an alternative realization.

6.3 Low-Complexity Design of a Single-Trit Multiplier

In order to reduce the transistor count, we study alternate ways of expressing the product and carry. In particular, we note that the product and carry can be rewritten in terms of the unary operators as Eqs. (6.3) and (6.4). In Eq. (6.3), we note that A corresponds to $0 \cdot A_0 + 1 \cdot A_1 + 2 \cdot A_2$. Further, the $+$ symbol in $(1 \cdot A_N^1 + A_1)$ corresponds to the logical OR (i.e., max) operation.

$$product = B_1 \cdot A + B_2 \cdot (1 \cdot A_N^1 + A_1) \qquad (6.3)$$
$$carry = B_2(1 \cdot A_N^1) \qquad (6.4)$$

A ternary multiplexer-based circuit based on Eqs. (6.3) and (6.4) is shown in Fig. 6.2. We note that $(1 \cdot A_N^1) + A_1$ need not be realized via independent circuits for A^1, A_1 etc. Instead, it can be realized as shown in Fig. 6.3. This circuit requires only three CNTFETs (namely, $T1$, $T2$ and $T3$) and operates as follows. The CNTFETs $T1$ and $T2$ have diameters of 1.487 nm while $T3$ has a diameter of 0.783 nm. Consider the input A = '0'. For this, the CNTFET $T2$ will be ON while the other two CNTFETs will be OFF in view of their threshold voltages. This leads the output to connect to ground, leading to logic '0'. Now, consider an input (for A) corresponding to logic '1'. CNTFET $T1$ will be ON while the other two CNTFETs will be OFF. This leads the output to connect to V_{DD}, which corresponds to logic '2'. Now consider the third possibility for the input A, namely logic 2'. In this case, the CNTFETs $T1$ and $T2$ are ON while the third CNTFET $T3$ is OFF. This results in the circuit working as a voltage divider which has an output of $\frac{V_{DD}}{2}$ corresponding to logic '1'.

Remark 6.2 The CNTFET count for this approach is 31. The details are as follows. The CNTFET implementation of $(1 \cdot A_N^1 + A_1)$ is shown in Fig. 6.3, which requires 3 CNTFETs and two more CNTFETs for the negative ternary inverter A_N. The other

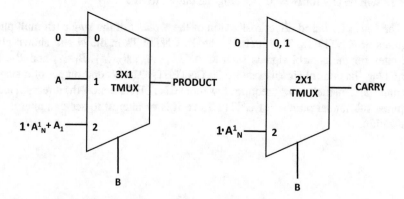

Fig. 6.2 Ternary multiplexer-based realization of a single trit multiplier

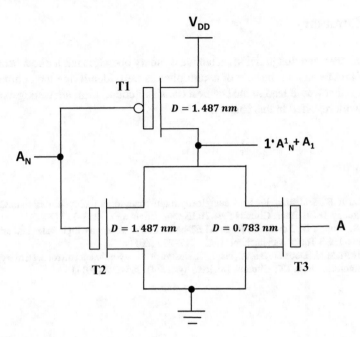

Fig. 6.3 CNTFET-based circuit for $(1 \cdot A_N^1) + A_1$

input to the carry multiplexer is $(1 \cdot A_N^1)$ which requires four CNTFETs. The 3×1 multiplexer takes 12 CNTFETs for the control signals and 6 CNTFETs for the transmission gate pairs. The 2×1 multiplexer requires 4 CNTFETs for the transmission gate pair. The same control signals can be used for both the multiplexers. Altogether, the multiplexer-based ternary single digit multiplier requires 31 CNTFETs.

Remark 6.3 It is worth noting that a single-trit multiplier has some similarities with a ternary half-adder: Both these units take two inputs (each 1-trit long) and produce two outputs (each 1-trit long). Further, they can also be realized using a similar approach. In particular, the approach based on one multiplexer for *sum* (in a ternary half-adder) extends to the *product* in a single-trit multiplier.

Remark 6.4 The designs in [2, 3] require 102 and 80 CNTFETs respectively. Hence, there is a substantial reduction in the CNTFET count using the approach investigated in this chapter.

6.4   Summary

We have observed that judicious selection of unary operators has a direct bearing on the CNTFET count in the case of a multiplier as well. Identifying the optimal combination (that would lead to the smallest transistor count) is an interesting extension to the work reported in this chapter.

References

1. Srinivasu, B., Sridharan, K.: Low-complexity multiternary digit multiplier design in CNTFET technology. IEEE Trans. Circuits Syst. II: Express Briefs **63**(8), 753–757 (2016)
2. Lin, S., Kim, Y.B., Lombardi, F.: CNTFET-based design of ternary logic gates and arithmetic circuits. IEEE Trans. Nanotechnol. **10**(2), 217–225 (2011)
3. Moaiyeri, M.H., Doostaregan, A., Navi, K.: Design of energy-efficient and robust ternary circuits for nanotechnology. IET Circuits, Devices, Syst. **5**(4), 285–296 (2011)

Chapter 7
Simulation Studies

In previous chapters, we have developed CNTFET-based designs for arithmetic operations such as addition and multiplication. In this chapter, we provide detailed simulation results for various circuits developed in the earlier chapters. The simulations use the MOSFET-like CNTFET model presented in [1, 2]. The parameters used for the simulation are described in Table 7.1.

SPICE programs for the simulations are presented in an appendix. The programs use the libraries described in [1, 3].

7.1 Simulation of Basic Ternary Logic Elements

7.1.1 Simulation of STI, PTI and NTI

The CNTFET-based simulation plot for the ternary inverter is shown in Fig. 7.1. This plot has all the three ternary inverters : standard ternary inverter, positive ternary inverter and negative ternary inverter. From the plot, the functionality of these can be observed (and is as expected).

7.1.2 Simulation of Other Unary Operators

The simulation plot for the cycle operators A^1 and A^2 for a given input A are shown in Fig. 7.2.

© Springer Nature Switzerland AG 2020
K. Sridharan et al., *Low-Complexity Arithmetic Circuit Design in Carbon Nanotube Field Effect Transistor Technology*, Carbon Nanostructures,
https://doi.org/10.1007/978-3-030-50699-5_7

Table 7.1 CNTFET parameters

Parameter	Description	Default value
Lg	Channel length	32 nm
Lsd	n+CNT source/drain full length from gate edge to S/D metal contact edge	16 nm
Lgef	Mean free path in intrinsic CNT	100 nm
Ccsd	The coupling capacitance between channel region and source/drain region.	0 F/m
Vfp	Flatband voltage of p-CNTFET	0 V
Vfn	Flatband voltage of n-CNTFET	0 V
Efo	The n+/p+ doped CNT fermi level	0.66 eV
nottube	Number of tubes on the CNTFET	1
Kox	The dielectric constant of high-k gate oxide material	16.0

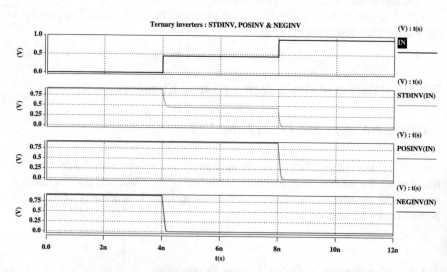

Fig. 7.1 Simulation for CNTFET-based ternary inverters: Here IN is the input while STDINV (IN), POSINV (IN) and NEGINV (IN) correspond to standard ternary inverter, positive ternary inverter and negative ternary inverter respectively

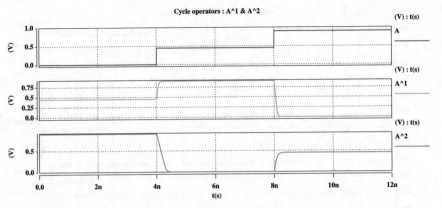

Fig. 7.2 Simulation for CNTFET-based ternary unary operators: Cycle operators A^1 and A^2. Here A is the input, $A^\wedge 1$ corresponds to A^1 and $A^\wedge 2$ corresponds to A^2

7.1.3 Simulation of Ternary NAND and NOR Gates

Figure 7.3 shows the CNTFET-based circuit simulation plot for the two-input ternary logic gates NAND and NOR. The plot has two inputs namely, A and B with two outputs given as NAND (A, B) and NOR (A, B) corresponding to the ternary NAND and NOR gates respectively.

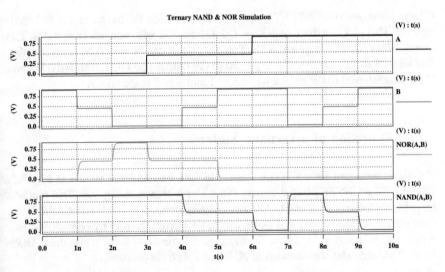

Fig. 7.3 Simulation for CNTFET-based two input ternary logic gates; A and B are the inputs while the plot of NAND (A, B) represents $\overline{A \cdot B}$ and NOR (A, B) represents $\overline{A + B}$

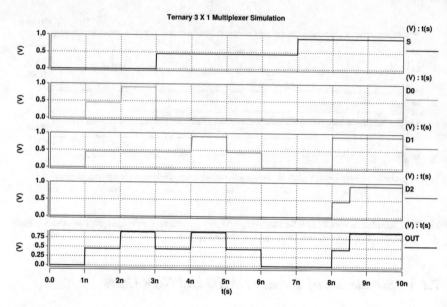

Fig. 7.4 Simulation for CNTFET-based ternary 3×1 multiplexer; S is the select line while D0, D1 and D2 are the data lines and OUT is the multiplexer output

7.1.4 Simulation of a Ternary 3 × 1 Multiplexer

Figure 7.4 shows the CNTFET-based circuit simulation for the ternary 3×1 multiplexer. This plot has three data lines D0, D1 and D2 with one select signal S. The output of the TMUX is given as OUT in the plot. It can be observed from the plot, that when S = '0', the data D0 is passed to the output line OUT. Similarly, D1 and D2 are transferred to the output when S = 1 and S = 2 respectively.

7.2 Simulation of a Ternary Adder

The simulation plot for a CNTFET-based ternary half-adder is shown in Fig. 7.5. The plot for a CNTFET-based ternary full adder is shown in Fig. 7.6. This plot has three inputs namely, A, B and C with two outputs, denoted by SUM and $CARRY$. The functionality of the ternary full adder can be seen from the plot by considering any input case. When A = '2' , B = '2' and C = '2', the SUM is '0' while the $CARRY$ is '2'. This indicates that the sum of A, B and C is 6 (in decimal).

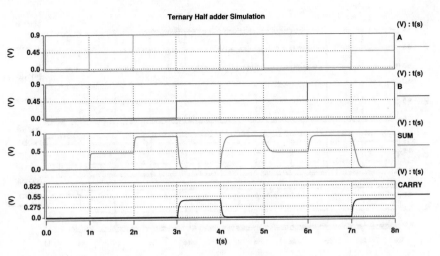

Fig. 7.5 Simulation for CNTFET-based ternary half adder; A and B are the ternary inputs while *SUM* and *CARRY* are the ternary outputs

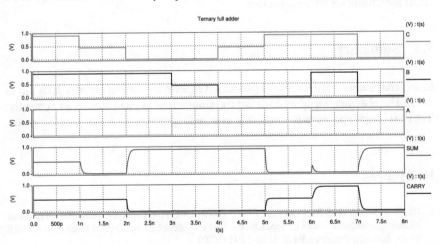

Fig. 7.6 Simulation for CNTFET-based ternary full adder; A,B and C are the ternary inputs while *SUM* and *CARRY* are the ternary outputs

7.3 Simulation of a Ternary Multiplier

The simulation plot for a CNTFET-based ternary multiplier is shown in Fig. 7.7.

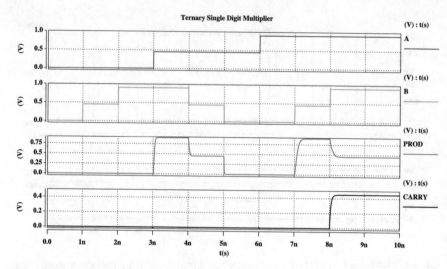

Fig. 7.7 Simulation for CNTFET-based ternary multiplier; *A* and *B* are the ternary inputs while PROD and CARRY are the ternary outputs

7.4 Summary

In this chapter, we have discussed simulation aspects for various circuits designed in the earlier chapters. Detailed code for the simulations is presented in an appendix. The next chapter presents an automatic synthesis approach for ternary logic circuits.

References

1. Deng, J., Wong, H.-S.P.: A compact SPICE model for carbon-nanotube field-effect transistors including nonidealities and its application—Part I: model of the intrinsic channel region. IEEE Trans. Electron Devices **54**(12), 3186–3194 (2007)
2. Stanford University CNTFET model, Website: Stanford University, Stanford, CA (2008). [Online] Available at http://nano.stanford.edu/model_stan_cnt.htm
3. Deng, J., Wong, H.S.P.: A compact SPICE model for carbon-nanotube field-effect transistors including nonidealities and its application—Part II: full device model and circuit performance benchmarking. IEEE Trans. Electron Devices **54**(12), 3195–3205 (2007)

Chapter 8
Automating the Synthesis Process

In previous chapters, we have developed CNTFET-based designs for specific operations such as addition and multiplication. We provided various "thumb-rules" for reducing the transistor count. In this chapter, we attempt to answer the following question: *Can the process of synthesis be automated ?* We provide an outline of an automatic synthesis procedure and also touch upon coding in Python. A study of the synthesis problem is also pursued in [1].

We use a geometrical approach and combine with unary operators discussed in the earlier chapters. In particular, the automation relies on a geometrical representation for three variables based on a ternary cube [2]. A computer program in Python for automatic synthesis is provided in an appendix.

8.1 Key Ideas

The main idea of automatic synthesis is to derive the ternary logic function as a sum-of-products expression consisting of unary operators. The approach is based on a few results which determine which way of representing a function is efficient. Consider Fig. 8.1 that corresponds to representation of a two-variable ternary function. The (actual) function to be realized can be represented by identifying one "common" signal which can serve as the select signal for a ternary multiplexer-based realization while the other signal/variable will correspond to the inputs for the multiplexer (and expressed in terms of unary operators). Note that this function can, in general, be expressed in more than one way and the task is to choose the one that yields less number of sum-of-product terms.

© Springer Nature Switzerland AG 2020

K. Sridharan et al., *Low-Complexity Arithmetic Circuit Design in Carbon Nanotube Field Effect Transistor Technology*, Carbon Nanostructures,
https://doi.org/10.1007/978-3-030-50699-5_8

Fig. 8.1 One layer (face) of
the cube (discussed earlier)
for handling two-variable
functions

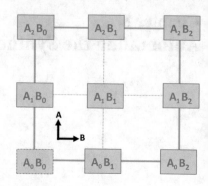

8.1.1   Choosing the Common Signal

We now illustrate the ideas via a two variable ternary function. Consider the function whose output is given in Table 8.1. We can directly write the expression from the truth table in terms of unary operators. There are two possibilities. One choice is to assume 'A' as the common signal. We can then write the function as $Y1 = A_0 \cdot 0 + A_1 \cdot 1 + A_2 \cdot 2 = A$ (Note that $A_0 = (2, 0, 0)$, $A_1 = (0, 2, 0)$ and $A_2 = (0, 0, 2)$). On the other hand, we could also consider 'A' as the unary operator (data for all the input lines for the corresponding multiplexer) and 'B' as the common signal. This results in $Y1 = B_0 \cdot A + B_1 \cdot A + B_2 \cdot A$. This, in turn, can be expressed as $A \cdot (B_0 + B_1 + B_2) = A \cdot 2 = A$. The first choice has fewer steps for minimizing the function compared to the latter case. Hence, the goal is to provide a method to choose the common signal which will lead to low complexity realization of the ternary logic functions.

This truth table can be represented in the cube as shown in Fig. 8.2. From the cube representation, when the elements along the horizontal row are 'similar', we

Table 8.1 Truth table of functions Y1 and Y2

A	B	Y1	Y2
0	0	0	0
0	1	0	1
0	2	0	2
1	0	1	0
1	1	1	1
1	2	1	2
2	0	2	0
2	1	2	1
2	2	2	2

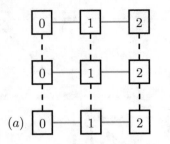 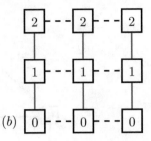

Fig. 8.2 (a) $Y1 = A$ and (b) $Y2 = B$

observe that the function can be written in terms of 'A' as $Y1 = A$. In the latter case, we can express in terms of 'B' as $Y2 = B$.

8.1.2 Choosing Unary Operators

The three elements in the corners of the cube represent the unary operator. The choice of unary operators depends on the way we scan the "lines". While selecting the axis for the unary operator, we seek answers to the following questions.

1. Does any horizontal or vertical line have zeros ?
2. Will (components along) any horizontal or vertical line be appropriate as input ?
3. If zeros are present in both horizontal and vertical directions, how do we proceed?
4. Are any two horizontal or vertical lines 'similar' ?

Remark 8.1 With respect to the third question above, we observe that we can compute the unary operators along the horizontal and vertical directions and choose the one with lower complexity.

8.2 Outline of An Algorithm for Automating the Synthesis Process

The algorithm primarily consists of two steps. Since the function can be general (that is, it can have n inputs), there is a need for decomposition. Hence, the first step is decomposition. This is followed by selection of the unary operator.

An n-input ternary logic function can have 3^n outputs. During the decomposition process, therefore an 'n-variable' cube is divided into blocks until we obtain 3^{n-2} two-variable blocks. For example, if we consider a three input block which has 27 outputs, this can be represented as the cube shown in Fig. 8.3. During the decomposition process, the 3-variable cube is divided into three two-variable cubes. The next step is choosing the unary operators from the two variable cubes.

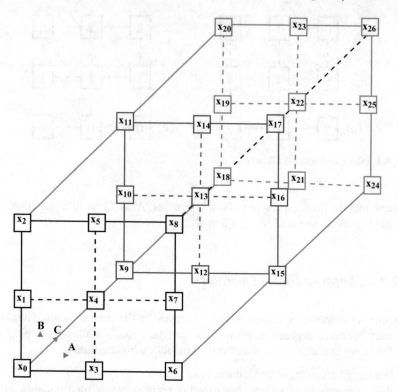

Fig. 8.3 Cube representation of a three variable function x_i, $i = 0, 1, 2, \ldots, 26$

Let us consider a three variable cube (shown in Fig. 8.3) as input to the decomposition algorithm. Note that this cube can be decomposed along any direction namely, A, B and C. The result of the decomposition will be three two-variable layers along $A = 0$, $A = 1$ and $A = 2$ (if we choose to decompose along A). Figures 8.4, 8.5 and

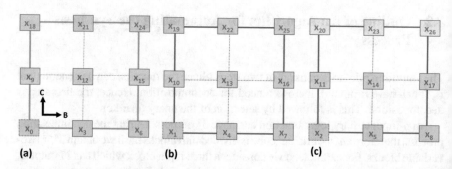

Fig. 8.4 Decomposition of a three-variable function into three layers of two-variable functions with A as common signal; **a** $A = 0$, **b** $A = 1$ and **c** $A = 2$

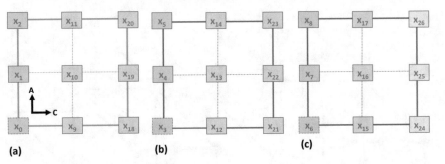

Fig. 8.5 Decomposition of a three-variable function into three layers of two-variable functions with B as common signal; **a** $B = 0$, **b** $B = 1$ and **c** $B = 2$

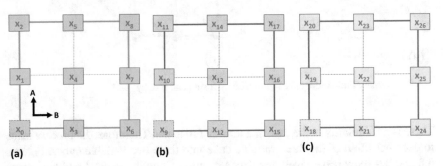

Fig. 8.6 Decomposition of a three-variable function into three layers of two-variable functions with C as common signal; **a** $C = 0$, **b** $C = 1$ and **c** $C = 2$

8.6 are the resulting two-variable layers along the directions A, B and C respectively. The main idea of the decomposition algorithm is to get simplified two-variable layers for a given n-variable input function. We next proceed to illustrate the entire process by considering an example of a ternary full-adder.

8.3 Illustration of the Automatic Synthesis Procedure for a Ternary Full Adder

We recall the truth table of the ternary full adder from earlier chapters. The truth table can be represented on a cube as shown in Fig. 8.7.

The decomposition algorithm starts by taking this cube as input. The algorithmic process first calculates the number of 0's, 1's and 2's in the truth table (and these are denoted by $count_0$, $count_1$ and $count_2$ respectively). If any of these counts is more than one third of the size of the truth table, then the decomposition algorithm tries to place all of them in a single two variable layer. In the case of ternary full adder, this does not apply. Hence, we proceed and look for similarity in the layers.

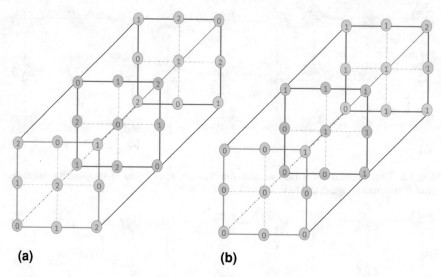

(a)   **(b)**

Fig. 8.7 Ternary Full-Adder cube representation **a** sum and **b** carry

There is no similarity in the case of a ternary full adder. Therefore, the process leads to decomposition of the three variable cube into three two-variable cubes. Figures 8.8 and 8.9 show respectively the decomposition along the vertical and horizontal directions (lines). Both of these result in the same number of two-variable layers. Hence, any one of these can be selected as the output of the decomposition process.

The second part of the automatic synthesis process involves selecting the appropriate unary operators from the decomposed two-variable layers. While selecting the unary operators, the algorithm considers the similarity in the operators and ignores the repeated operators or operators such as constant ('000') and buffer ('012'). The outcome of these steps in the synthesis algorithm is as follows.

```
The input to the automatic synthesis algorithm is as follows :
no. of inputs = 3
no. of outputs =2
Function : SUM
Results are as follows::
('Actual number of two variable layers :   6)

Decomposed layers   ::
For first output, CARRY :
[0, 0, 0, 0, 0, 1, 0, 1, 1]
[0, 0, 1, 0, 1, 1, 1, 1, 1]
[0, 1, 1, 1, 1, 1, 1, 1, 2]
For second output, SUM :
[0, 1, 2, 1, 2, 0, 2, 0, 1]
```

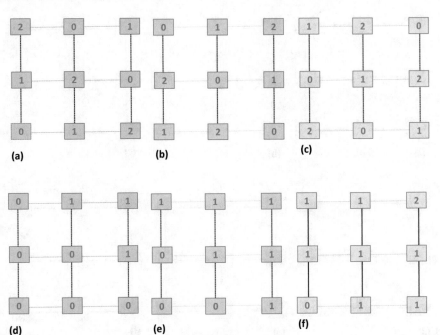

Fig. 8.8 Decomposition output of a ternary full adder along the vertical lines; SUM: **a** B = 0 **b** B = 1 **c** B = 2 and CARRY:**d** B = 0 **e** B = 1 **f** B = 2

```
[1, 2, 0, 2, 0, 1, 0, 1, 2]
[2, 0, 1, 0, 1, 2, 1, 2, 0]

('Total number of two variable layers after decomposition :   6)

Unary operators :
['000', '001', '011', '111', '112', '012', '120', '201']
Total number of unary operators : 8
No. of unary operators required : 6
Number of CNTFETs for unary operators : 29
Number of MUXes will be : 8
Total number of CNTFETS : 105
```

8.4 Python Code for Automatic Synthesis

An appendix containing the Python code developed for automatic synthesis is included in this book. The appendix also gives details of a sample run.

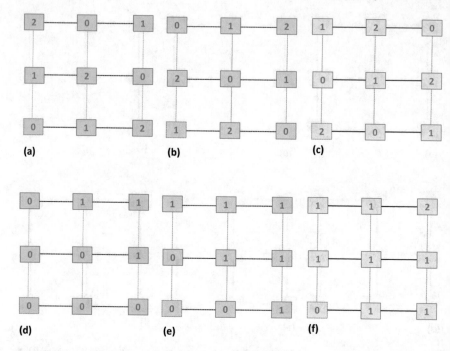

Fig. 8.9 Decomposition output of a ternary full adder along the horizontal lines; SUM: **a** A = 0 **b** A = 1 **c** A = 2 and CARRY:**d** A = 0 **e** A = 1 **f** A = 2

8.5   Summary

The focus of this chapter has been on automating the synthesis process. The proposed approach lends itself readily to adaptation to various device technologies.

References

1. Srinivasu, B., Sridharan, K.: A synthesis methodology for ternary logic circuits in emerging device technologies. IEEE Trans. Circuits Syst. I: Regul. Pap. **64**(8), 2146–2159 (2017)
2. Hurst, S.L.: An extension of binary minimization techniques to ternary equations. Comput. J. **11**(3), 277–286 (1968)

Chapter 9
The Road Ahead

This research has studied digital design in the context of emerging nanotechnologies. In particular, we have studied the problem of designing arithmetic circuits in Carbon Nanotube Field Effect Transistor technology. We have presented a number of theoretical results on ternary logic. The results facilitate reduction of transistor count for various circuits. We now list the contributions and touch upon extensions to the work.

9.1 Contributions of This Research

The contributions of this book are as follows.

- An introduction to carbon nanotubes, CNTFETs and ternary logic.
- Realization of basic logic elements using CNTFETs.
- Efficient CNTFET design of single-trit adders including half-adders and full-adders.
- Low-complexity CNTFET-based design of a multi-trit adder.
- A low-complexity CNTFET-based multiplier.
- Simulations of various designs.
- A general approach to automate the synthesis process.
- HSPICE and Python code for CNTFET-based design and simulation.

© Springer Nature Switzerland AG 2020
K. Sridharan et al., *Low-Complexity Arithmetic Circuit Design in Carbon Nanotube Field Effect Transistor Technology*, Carbon Nanostructures,
https://doi.org/10.1007/978-3-030-50699-5_9

9.2   Extensions

9.2.1   Designs with Optimal Number of Transistors

This book has discussed unary operators of multi-valued logic and shown how one can design CNTFET-based circuits with low transistor count for realizing various arithmetic circuits. Multiplexer-based designs in the ternary setting have been extensively studied. Alternatives to multiplexer-based designs have also been explored. However, there is room for additional work pertaining to optimization of the transistor count.

9.2.2   Alternate Approaches for Synthesis

This book has largely concentrated on the use of 3:1 multiplexers for CNTFET-based realization. Recently, a 2:1 multiplexer-based approach has been proposed for synthesis in [1] using ternary transformed binary decision diagrams. It would be of interest to study different synthesis approaches further from the point of view of development of computer-aided design tools for CNTFET-based circuits.

9.2.3   Extension To Quaternary Logic

The focus of this book has been on ternary logic circuits in CNTFET technology. Other multi-valued logic systems may be explored. Recently, quaternary full-adder designs have been reported [2, 3]. Much more can be done in this direction including efficient design of multi-digit adders and multipliers.

9.2.4   Studies on Crosstalk Mitigation

This book focusses on design of circuits with low transistor count. It is well known that the noise margin is inherently lower in multivalued logic systems. As a result, crosstalk handling is a non-trivial problem. Recently, the authors in [4] have proposed an approach for active shielding of multiwalled carbon nanotube interconnects to eliminate crosstalk-induced failures in ternary logic. Further investigations on this subject would be valuable.

9.2.5 Applications of the Work to Other Nanotechnologies

The proposed ternary logic-based arithmetic circuit designs can also be applied to other devices that support multiple-valued logic. One example of this is the Quantum Dot Gate FET (QDGFET) [5]. In Quantum Dot Gate FET, a quantum dot layer is imposed on top of the gate insulator in a conventional MOSFET. The presence of the quantum dot layer allows three states with one intermediate state in addition to the usual ON and OFF states. The threshold voltage of the device can be changed by changing the insulator thickness or by changing the number of quantum dot layers.

9.2.6 Fabrication and Implementation of CNTFET-Based Arithmetic Circuits

Another area of work for further exploration is fabrication of the circuits proposed in this book. The state of the art in fabrication is a 16-bit microprocessor based on complementary carbon nanotube transistors [6]. Currently, there does not appear to be any chip that implements CNTFET-based ternary logic circuits. It would be of interest to examine the fabrication-level challenges for arithmetic circuits.

9.2.7 Application to Computations for High-Level Tasks

One can also take the designs for arithmetic circuits a step further. In particular, it would be of interest to design low-complexity CNTFET-based circuits for matrix multiplication and computation of various discrete orthogonal transforms.

9.3 Concluding Remarks

Research on ternary function minimization has been pursued as early as 1968. However, the non-availability of appropriate devices to realize ternary and other multi-valued logic systems has for long remained a concern but the emergence of several new device technologies has led to renewed interest in ternary and quaternary logic in particular. We believe that, in addition to theoretical development, advanced tools for automated optimization as well as design in upcoming nanotechnologies will emerge in the years to come.

References

1. Vudadha, C., Surya, A., Agrawal, S., Srinivas, M.B.: Synthesis of ternary logic circuits using 2:1 multiplexers. IEEE Trans. Circuits Syst.-I: Regul. Pap. **65**(12), 4313–4325 (2018)
2. Daraei, A., Hosseini, S.A.: Novel energy-efficient and high-noise margin quaternary circuits in nanoelectronics. Int. J. Electron. Commun. (AEÜ)
3. Patel, P., Doddapaneni, N., Gadgil, S., Vudadha, C.: Design of area optimised, energy efficient quaternary circuits using CNTFETs. In: Proceedings of 2019 IEEE International Symposium on Smart Electronic Systems, pp. 280–283 (2019)
4. Khezeli, M.R., Moaiyeri, M.H., Jalali, A.: Active shielding of MWCNT bundle interconnets: an efficient approach to cancellation of crosstalk-induced functional failures in ternary logic. IEEE Trans. Electromag. Compat. **61**(1), 100–110 (2019)
5. Karmakar, S., Chandy, J.A., Jain, F.C.: Design of ternary logic combinational circuits based on quantum dot gate FETs. IEEE Trans. VLSI Syst. **21**(5), 793–806 (2013)
6. Hills, G., Lau, C., Wright, A., Fuller, S., Bishop, M.D., Srimani, T., Kanhaiya, P., Ho, R., Amer, A., Stein, Y., Murphy, D., Arvind, Chandrakasan, A., Shulaker, M.M.: Modern microprocessor built from complementary carbon nanotube transistors. Nature **572**, 595–602 (2019)

Appendix A
HSPICE Code for Various Ternary Logic Circuits

In this appendix, we present HSPICE code developed for simulation of various designs. We begin with code that is common for all circuits. We then present code that is specific to each logic element.

A.1 Common HSPICE Code for a CNTFET Circuit

```
.TITLE 'CNTFET-based circuit simulation'
*********************************************************
*Options for optimal accuracy, convergence, and runtime
*********************************************************
.options POST
.options AUTOSTOP
.options INGOLD=2      DCON=1
.options GSHUNT=1e-12 RMIN=1e-15
.options ABSTOL=1e-5  ABSVDC=1e-4
.options RELTOL=1e-2  RELVDC=1e-2
.options NUMDGT=4      PIVOT=13
.param   TEMP=27
*********************************************************
*********************************************************
**** Include CNFET.lib in the same folder *****

.lib 'CNFET.lib' CNFET
*********************************************************
*Supply and voltage parameters**
*********************************************************
.param Supply=0.9
```

© Springer Nature Switzerland AG 2020

K. Sridharan et al., *Low-Complexity Arithmetic Circuit Design in Carbon Nanotube Field Effect Transistor Technology*, Carbon Nanostructures,
https://doi.org/10.1007/978-3-030-50699-5

```
.param Vg='Supply'
.param Vd='Supply'
.param Vm=0.45


***************************************************
*Key transistor parameters****
***************************************************
.param Ccsd=0      CoupleRatio=0
.param m_cnt=1     Efo=0.6
.param Wg=0        Cb=40e-12
.param Lg=32e-9    Lgef=100e-9
.param Vfn=0       Vfp=0
.param m=19        n=0
.param Hox=4e-9    Kox=16
.param nottube=1


***************************************************
* Power supply definition
***************************************************
Vmm       Vmm        Gnd        Vm
Vdd       Vdd        Gnd        Vd
Vss       Source     Gnd        0
Vsub      Sub        Gnd        0


.PARAM T=1n
```

A.2   HSPICE Code for a Negative Ternary Inverter (NTI)

In addition to the common code in Sect. A.1, the following is required for simulation
of an NTI.

```
*****************************************
*** Specification of three-valued input
** The input IN is 0 V upto 2 nsec,
**              IN is 0.45 V from '2+0.01p' to 4 nsec
**              IN is 0.9 V from '4+0.01p' to 6 nsec
*****************************************
VIN IN Gnd PWL(0.0 0.0 '2*T' 0.0 '2*T+0.01p' 0.45
+ '4*T' 0.45 '4*T+0.01p' 0.9 '6*T' 0.9)



*****************************************
```

```
* Negative NOT Gate Subcircuit module definition
*************************************************

.SUBCKT NEGINV IN OUT Source Sub Vdd

XCNT1 OUT IN Vdd Vdd PCNFET Lch=Lg  Lgeff='Lgef' Lss=32e-9
+ Ldd=32e-9  Kgate='Kox' Tox='Hox' Csub='Cb' Vfbn='Vfn' Dout=0
+ Sout=0  Pitch=20e-9  n1=10  n2=0  tubes=nottube

XCNT2 OUT IN Source Sub NCNFET Lch=Lg  Lgeff='Lgef' Lss=32e-9
+ Ldd=32e-9  Kgate='Kox' Tox='Hox' Csub='Cb' Vfbn='Vfn' Dout=0
+ Sout=0  Pitch=20e-9  n1=19  n2=n  tubes=nottube

C1 OUT Gnd 0.5f

.ENDS NEGINV

*****************************************
*** Next give command for simulating the NTI
******************************************

XCKT IN OUT Source Sub Vdd NEGINV

*****************************************
*** Then give command for transient analysis of the circuit
*****************************************
.tran 100p '6*T'

*****************************************
*** Give the following command for calculating
*** the average and peak power consumption
*****************************************

.MEAS TRAN AVGPOW AVG POWER FROM=0n TO='6*T'
.MEAS TRAN PEAKPOWER MAX POWER FROM=0n TO='6*T'


.end
```

A.3   HSPICE Codes for Other Unary Circuits

In addition to the code given in Sect. A.1, the following is required to simulate various unary operator circuits. The following code for input applies to STI, PTI and other unary operators.

```
********************************************
*** Specification of three-valued input
** The input IN is 0 V upto 2 nsec,
**              IN is 0.45 V from '2+0.01p' to 4 nsec
**              IN is 0.9 V from '4+0.01p' to 6 nsec
********************************************
VIN IN Gnd PWL(0.0 0.0 '2*T' 0.0 '2*T+0.01p' 0.45
+ '4*T' 0.45 '4*T+0.01p' 0.9 '6*T' 0.9)
```

A.3.1   Standard Ternary Inverter

In addition to the common code in Sect. A.1 and the input description in Sect. A.3, the following code is required. Note that $OUT = \overline{IN}$ in the code below.

```
.SUBCKT STDINV IN OUT Source Sub Vdd

XCNT1 N2 IN Vdd Vdd PCNFET Lch=Lg  Lgeff='Lgef' Lss=32e-9
+ Ldd=32e-9  Kgate='Kox' Tox='Hox' Csub='Cb' Vfbn='Vfn'
+ Dout=0  Sout=0 Pitch=20e-9 n1=19  n2=0  tubes=nottube

XCNT2 OUT IN Vdd Vdd PCNFET Lch=Lg  Lgeff='Lgef' Lss=32e-9
+ Ldd=32e-9  Kgate='Kox' Tox='Hox' Csub='Cb' Vfbn='Vfn'
+ Dout=0  Sout=0  Pitch=20e-9 n1=10  n2=0  tubes=nottube

XCNT4 N1 IN Source Sub NCNFET Lch=Lg  Lgeff='Lgef' Lss=32e-9
+ Ldd=32e-9  Kgate='Kox' Tox='Hox' Csub='Cb' Vfbn='Vfn'
+ Dout=0  Sout=0  Pitch=20e-9 n1=19  n2=0  tubes=nottube

XCNT5 OUT IN Source Sub NCNFET Lch=Lg  Lgeff='Lgef' Lss=32e-9
+ Ldd=32e-9  Kgate='Kox' Tox='Hox' Csub='Cb' Vfbn='Vfn'
+ Dout=0  Sout=0 Pitch=20e-9  n1=10  n2=0  tubes=nottube

XCNT3 OUT OUT N2 N2 PCNFET Lch=Lg  Lgeff='Lgef' Lss=32e-9
```

```
+ Ldd=32e-9  Kgate='Kox'  Tox='Hox'  Csub='Cb'  Vfbn='Vfn'
+ Dout=0 Sout=0  Pitch=20e-9  n1=13  n2=0  tubes=nottube

XCNT6 OUT OUT N1 N1 NCNFET Lch=Lg  Lgeff='Lgef' Lss=32e-9
+ Ldd=32e-9  Kgate='Kox'  Tox='Hox'  Csub='Cb'  Vfbn='Vfn'
+ Dout=0 Sout=0  Pitch=20e-9  n1=13  n2=0  tubes=nottube


C1 OUT Gnd 0.5f

.ENDS STDINV
```

A.3.2   Positive Ternary Inverter

In addition to the common code in Sect. A.1 and the input description in Sect. A.3, the following is required.

```
.SUBCKT POSINV IN OUT Source Sub Vdd

XCNT1 OUT IN Vdd Vdd PCNFET Lch=Lg  Lgeff='Lgef' Lss=32e-9
+ Ldd=32e-9  Kgate='Kox'  Tox='Hox'  Csub='Cb'  Vfbn='Vfn' Dout=0
+ Sout=0  Pitch=20e-9  n1=19  n2=0  tubes=nottube

XCNT2 OUT IN Source Sub NCNFET Lch=Lg  Lgeff='Lgef' Lss=32e-9
+ Ldd=32e-9  Kgate='Kox'  Tox='Hox'  Csub='Cb'  Vfbn='Vfn' Dout=0
+ Sout=0  Pitch=20e-9  n1=10  n2=0  tubes=nottube

C1 OUT Gnd 0.5f

.ENDS POSINV
```

A.3.3   Negative Ternary Inverter

In addition to the common code in Sect. A.1 and the input description in Sect. A.3, the following is required.

```
.SUBCKT NEGINV IN OUT Source Sub Vdd

XCNT1 OUT IN Vdd Vdd PCNFET Lch=Lg  Lgeff='Lgef' Lss=32e-9
```

```
+ Ldd=32e-9   Kgate='Kox'  Tox='Hox'  Csub='Cb'  Vfbn='Vfn'  Dout=0
+ Sout=0   Pitch=20e-9   n1=10   n2=0   tubes=nottube

XCNT2 OUT IN Source Sub NCNFET Lch=Lg  Lgeff='Lgef' Lss=32e-9
+ Ldd=32e-9   Kgate='Kox'  Tox='Hox'  Csub='Cb'  Vfbn='Vfn'  Dout=0
+ Sout=0   Pitch=20e-9   n1=19   n2=0   tubes=nottube

C1 OUT Gnd 0.5f

.ENDS NEGINV
```

A.3.4   Binary Inverter

In addition to the common code in Sect. A.1 and the input description in Sect. A.3, the following is required. Note that $OUT = \overline{IN}$ where IN and OUT are binary signals.

```
.SUBCKT INV IN OUT Source Sub Vdd

XCNT1 OUT IN Vdd Vdd PCNFET Lch=Lg  Lgeff='Lgef' Lss=32e-9
+ Ldd=32e-9   Kgate='Kox'  Tox='Hox'  Csub='Cb'  Vfbn='Vfn'  Dout=0
+ Sout=0   Pitch=20e-9   n1=19   n2=0   tubes=nottube

XCNT2 OUT IN Source Sub NCNFET Lch=Lg  Lgeff='Lgef' Lss=32e-9
+ Ldd=32e-9   Kgate='Kox'  Tox='Hox'  Csub='Cb'  Vfbn='Vfn'  Dout=0
+ Sout=0   Pitch=20e-9   n1=19   n2=0   tubes=nottube

C1 OUT Gnd 0.5f

.ENDS INV
```

A.3.5   Cycle Operators

In addition to the common code in Sect. A.1 and the input description in Sect. A.3, the following is required. Note that $OUT = A^1$ in the code below.

```
.SUBCKT cycle_A1 IN OUT Source Sub Vdd

XNEGINV IN X Source Sub Vdd NEGINV
```

```
XCNT1 OUT IN Vdd Vdd PCNFET Lch=Lg  Lgeff='Lgef' Lss=32e-9
+ Ldd=32e-9  Kgate='Kox' Tox='Hox' Csub='Cb' Vfbn='Vfn' Dout=0
+ Sout=0  Pitch=20e-9  n1=19 n2=0  tubes=nottube


XCNT2 OUT IN Gnd Gnd NCNFET Lch=Lg  Lgeff='Lgef' Lss=32e-9
+ Ldd=32e-9  Kgate='Kox' Tox='Hox' Csub='Cb' Vfbn='Vfn' Dout=0
+ Sout=0  Pitch=20e-9  n1=10  n2=0  tubes=nottube


XCNT3 OUT X Gnd Gnd NCNFET Lch=Lg  Lgeff='Lgef' Lss=32e-9
+ Ldd=32e-9  Kgate='Kox' Tox='Hox' Csub='Cb' Vfbn='Vfn' Dout=0
+ Sout=0  Pitch=20e-9  n1=19  n2=0  tubes=nottube


C1 OUT Gnd 0.5f


.ENDS cycle_A1
```

The code for A^1 is complete. We next give the code for A^2. In the code below, $OUT = A^2$.

```
.SUBCKT cycle_A2 IN OUT Source Sub Vdd

XPOSINV IN X Source Sub Vdd POSINV

XCNT1 OUT IN Vdd Vdd PCNFET Lch=Lg  Lgeff='Lgef' Lss=32e-9
+ Ldd=32e-9  Kgate='Kox' Tox='Hox' Csub='Cb' Vfbn='Vfn' Dout=0
+ Sout=0  Pitch=20e-9  n1=10 n2=0  tubes=nottube


XCNT2 OUT X Vdd Vdd PCNFET Lch=Lg  Lgeff='Lgef' Lss=32e-9
+ Ldd=32e-9  Kgate='Kox' Tox='Hox' Csub='Cb' Vfbn='Vfn' Dout=0
+ Sout=0  Pitch=20e-9  n1=19 n2=0  tubes=nottube


XCNT3 OUT IN Gnd Gnd NCNFET Lch=Lg  Lgeff='Lgef' Lss=32e-9
+ Ldd=32e-9  Kgate='Kox' Tox='Hox' Csub='Cb' Vfbn='Vfn' Dout=0
+ Sout=0  Pitch=20e-9  n1=19  n2=0  tubes=nottube


C1 OUT Gnd 0.5f


.ENDS cycle_A2
```

A.3.6 Decisive Operators

In addition to the common code in Sect. A.1 and the input description in Sect. A.3, the following is required. Note that $OUT = A_0$ in the code below.

```
.SUBCKT Dec_A0 IN OUT Source Sub Vdd

XCNT1 OUT IN Vdd Vdd PCNFET Lch=Lg  Lgeff='Lgef' Lss=32e-9
+ Ldd=32e-9  Kgate='Kox' Tox='Hox' Csub='Cb' Vfbn='Vfn' Dout=0
+ Sout=0  Pitch=20e-9  n1=10  n2=0  tubes=nottube

XCNT2 OUT IN Source Sub NCNFET Lch=Lg  Lgeff='Lgef' Lss=32e-9
+ Ldd=32e-9  Kgate='Kox' Tox='Hox' Csub='Cb' Vfbn='Vfn' Dout=0
+ Sout=0  Pitch=20e-9  n1=19  n2=0  tubes=nottube

C1 OUT Gnd 0.5f

.ENDS Dec_A0
```

The code for A_0 is complete. We now begin code for A_1. Note that $OUT = A_1$ in the code below.

```
.SUBCKT Dec_A1 IN OUT Source Sub Vdd

XNEGINV0 IN X Source Sub Vdd NEGINV

XCNT1 N2 IN Vdd Vdd PCNFET Lch=Lg  Lgeff='Lgef' Lss=32e-9
+ Ldd=32e-9  Kgate='Kox' Tox='Hox' Csub='Cb' Vfbn='Vfn' Dout=0
+ Sout=0  Pitch=20e-9  n1=19  n2=0  tubes=nottube

XCNT2 OUT X N2 N2 PCNFET Lch=Lg  Lgeff='Lgef' Lss=32e-9
+ Ldd=32e-9  Kgate='Kox' Tox='Hox' Csub='Cb' Vfbn='Vfn' Dout=0
+ Sout=0  Pitch=20e-9  n1=19  n2=0  tubes=nottube

XCNT3 OUT IN Source Sub NCNFET Lch=Lg  Lgeff='Lgef' Lss=32e-9
+ Ldd=32e-9  Kgate='Kox' Tox='Hox' Csub='Cb' Vfbn='Vfn' Dout=0
+ Sout=0  Pitch=20e-9  n1=10  n2=0  tubes=nottube

XCNT4 OUT X N1 N1 NCNFET Lch=Lg  Lgeff='Lgef' Lss=32e-9
+ Ldd=32e-9  Kgate='Kox' Tox='Hox' Csub='Cb' Vfbn='Vfn' Dout=0
+ Sout=0  Pitch=20e-9  n1=19  n2=0  tubes=nottube
```

```
XCNT5 N1 X Source Sub NCNFET Lch=Lg  Lgeff='Lgef' Lss=32e-9
+ Ldd=32e-9  Kgate='Kox' Tox='Hox' Csub='Cb' Vfbn='Vfn' Dout=0
+ Sout=0  Pitch=20e-9  n1=19  n2=0  tubes=nottube

C1 OUT Gnd 0.5f

.ENDS Dec_A1
```

The code for A_1 is complete. We now begin code for A_2. Note that $OUT = A^2$ in the code below.

```
.SUBCKT Dec_A2 IN OUT Source Sub Vdd

XCNT1 Y0 IN Vdd Vdd PCNFET Lch=Lg  Lgeff='Lgef' Lss=32e-9
+ Ldd=32e-9  Kgate='Kox' Tox='Hox' Csub='Cb' Vfbn='Vfn' Dout=0
+ Sout=0  Pitch=20e-9  n1=19  n2=0  tubes=nottube

XCNT2 Y0 IN Source Sub NCNFET Lch=Lg  Lgeff='Lgef' Lss=32e-9
+ Ldd=32e-9  Kgate='Kox' Tox='Hox' Csub='Cb' Vfbn='Vfn' Dout=0
+ Sout=0  Pitch=20e-9  n1=10  n2=0  tubes=nottube

XINV Y0 OUT Source Sub Vdd INV

C1 OUT Gnd 0.5f

.ENDS Dec_A2
```

A.3.7   HSPICE Code for Other Important Unary Operators

In addition to the common code in Sect. A.1 and the input description in Sect. A.3, the following is required. Note that $OUT = 1 \cdot \overline{A^1}$, in the code below (and the inverter corresponds to an NTI).

```
.SUBCKT CKT_001 IN OUT Source Sub Vdd

XPOSINV IN INBAR Source Sub Vdd POSINV

XCNT1 OUT INBAR Vdd Vdd PCNFET Lch=Lg  Lgeff='Lgef' Lss=32e-9
```

```
+ Ldd=32e-9  Kgate='Kox'  Tox='Hox'  Csub='Cb'  Vfbn='Vfn'  Dout=0
+ Sout=0  Pitch=20e-9  n1=19  n2=0  tubes=nottube

XCNT2 OUT Vdd Source Sub NCNFET Lch=Lg  Lgeff='Lgef'  Lss=32e-9
+ Ldd=32e-9  Kgate='Kox'  Tox='Hox'  Csub='Cb'  Vfbn='Vfn'  Dout=0
+ Sout=0  Pitch=20e-9  n1=19  n2=0  tubes=nottube

C1 OUT Gnd 0.5f

.ENDS CKT_001
```

We are done with the code for $1 \cdot \overline{A^1}$. We now begin the code for $1 \cdot \overline{A^2}$. Note that $OUT = 1 \cdot \overline{A^2}$ where the inverter is a PTI.

```
.SUBCKT CKT_011 IN OUT Source Sub Vdd

XNEGINV IN INBAR Source Sub Vdd NEGINV

XCNT1 OUT INBAR Vdd Vdd PCNFET Lch=Lg  Lgeff='Lgef'  Lss=32e-9
+ Ldd=32e-9  Kgate='Kox'  Tox='Hox'  Csub='Cb'  Vfbn='Vfn'  Dout=0
+ Sout=0  Pitch=20e-9  n1=19  n2=0  tubes=nottube

XCNT2 OUT Vdd Source Sub NCNFET Lch=Lg  Lgeff='Lgef'  Lss=32e-9
+ Ldd=32e-9  Kgate='Kox'  Tox='Hox'  Csub='Cb'  Vfbn='Vfn'  Dout=0
+ Sout=0  Pitch=20e-9  n1=19  n2=0  tubes=nottube

C1 OUT Gnd 0.5f

.ENDS CKT_011
```

We are done with the code for $1 \cdot \overline{A^2}$. We now begin the code for $1 + \overline{A^1}$. Note that $OUT = 1 + \overline{A^1}$ in the code below where the inverter is a negative ternary inverter.

```
.SUBCKT CKT_112 IN  OUT Source Sub Vdd

XPOSINV IN X Source Sub Vdd POSINV

XINV X XBAR Source Sub Vdd INV

XCNT1 OUT XBAR Vdd Vdd PCNFET Lch=Lg  Lgeff='Lgef'  Lss=32e-9
```

```
+ Ldd=32e-9  Kgate='Kox' Tox='Hox' Csub='Cb' Vfbn='Vfn' Dout=0
+ Sout=0  Pitch=20e-9  n1=19  n2=0  tubes=nottube

XCNT2 OUT X Vdd Vdd PCNFET Lch=Lg  Lgeff='Lgef' Lss=32e-9
+ Ldd=32e-9  Kgate='Kox' Tox='Hox' Csub='Cb' Vfbn='Vfn' Dout=0
+ Sout=0  Pitch=20e-9  n1=19  n2=0  tubes=nottube

XCNT3 OUT X Source Sub NCNFET Lch=Lg  Lgeff='Lgef' Lss=32e-9
+ Ldd=32e-9  Kgate='Kox' Tox='Hox' Csub='Cb' Vfbn='Vfn' Dout=0
+ Sout=0  Pitch=20e-9  n1=19  n2=0  tubes=nottube

C1 OUT Gnd 0.5f

.ENDS CKT_112
```

We are done with the code for $1 + \overline{A^1}$. We now present the code for $A_1 + (1 \cdot \overline{A^1})$. Note that $OUT = A_1 + (1 \cdot \overline{A^1})$ where the inverter is an NTI.

```
.SUBCKT CKT_021 IN  OUT Source Sub Vdd

XNEGINV IN X Source Sub Vdd NEGINV


XCNT1 OUT R Vdd Vdd PCNFET Lch=Lg  Lgeff='Lgef' Lss=32e-9
+ Ldd=32e-9  Kgate='Kox' Tox='Hox' Csub='Cb' Vfbn='Vfn' Dout=0
+ Sout=0  Pitch=20e-9  n1=19  n2=0  tubes=nottube

XCNT2 OUT X Source Sub NCNFET Lch=Lg  Lgeff='Lgef' Lss=32e-9
+ Ldd=32e-9  Kgate='Kox' Tox='Hox' Csub='Cb' Vfbn='Vfn' Dout=0
+ Sout=0  Pitch=20e-9  n1=19  n2=0  tubes=nottube

XCNT3 OUT IN Source Sub NCNFET Lch=Lg  Lgeff='Lgef' Lss=32e-9
+ Ldd=32e-9  Kgate='Kox' Tox='Hox' Csub='Cb' Vfbn='Vfn' Dout=0
+ Sout=0  Pitch=20e-9  n1=10  n2=0  tubes=nottube

C1 OUT Gnd 0.5f

.ENDS CKT_021
```

A.4   HSPICE Codes for Two Input Circuits

The inputs A and B for all two-input circuits is specified as shown below.

```
VA A Gnd PWL(0.0 0.0 '3*T' 0 '3*T+0.01p' 0.45 '6*T' 0.45
+ '6*T+0.01p' 0.9 '9*T' 0.9 '9*T+0.01p') **0.0 '9*T' 0.0
+ '9*T+0.01p' 0.9 '10*T' 0.9)


VB B Gnd PWL(0.0 0.9 '3*T' 0.9 '3*T+0.01p' 0.45 '4*T' 0.45
+ '4*T+0.01p' 0.0 '6*T' 0.0 '6*T+0.01p' 0.9 '7*T' 0.9
+ '7*T+0.01p' 0.0 '8*T' 0.0 '8*T+0.01p' 0.9 '10*T' 0.9)
```

A.4.1   Two Input Ternary NAND Gate

In the code below, $OUT = \overline{A \cdot B}$.

```
.SUBCKT TNAND A B OUT Source Sub Vdd

XCNT1 N1 A Vdd Vdd PCNFET Lch=Lg  Lgeff='Lgef' Lss=32e-9
+ Ldd=32e-9  Kgate='Kox' Tox='Hox' Csub='Cb' Vfbn='Vfn' Dout=0
+ Sout=0  Pitch=20e-9  n1=19  n2=0  tubes=nottube

XCNT2 N1 B Vdd Vdd PCNFET Lch=Lg  Lgeff='Lgef' Lss=32e-9
+ Ldd=32e-9  Kgate='Kox' Tox='Hox' Csub='Cb' Vfbn='Vfn' Dout=0
+ Sout=0  Pitch=20e-9  n1=19  n2=0  tubes=nottube

XCNT3 OUT A Vdd Vdd PCNFET Lch=Lg  Lgeff='Lgef' Lss=32e-9
+ Ldd=32e-9  Kgate='Kox' Tox='Hox' Csub='Cb' Vfbn='Vfn' Dout=0
+ Sout=0  Pitch=20e-9  n1=10  n2=0  tubes=nottube

XCNT4 OUT B Vdd Vdd PCNFET Lch=Lg  Lgeff='Lgef' Lss=32e-9
+ Ldd=32e-9  Kgate='Kox' Tox='Hox' Csub='Cb' Vfbn='Vfn' Dout=0
+ Sout=0  Pitch=20e-9  n1=10  n2=0  tubes=nottube

XCNT5 OUT OUT N1 N1 PCNFET Lch=Lg  Lgeff='Lgef' Lss=32e-9
+ Ldd=32e-9  Kgate='Kox' Tox='Hox' Csub='Cb' Vfbn='Vfn' Dout=0
+ Sout=0  Pitch=20e-9  n1=13  n2=0  tubes=nottube

XCNT6 OUT OUT N2 N2 NCNFET Lch=Lg  Lgeff='Lgef' Lss=32e-9
```

```
+ Ldd=32e-9  Kgate='Kox' Tox='Hox' Csub='Cb' Vfbn='Vfn' Dout=0
+ Sout=0  Pitch=20e-9  n1=13  n2=0  tubes=nottube

XCNT7 N2 B N3 N3 NCNFET Lch=Lg  Lgeff='Lgef' Lss=32e-9
+ Ldd=32e-9  Kgate='Kox' Tox='Hox' Csub='Cb' Vfbn='Vfn' Dout=0
+ Sout=0  Pitch=20e-9  n1=19  n2=0  tubes=nottube

XCNT8 N3 A Source Sub NCNFET Lch=Lg  Lgeff='Lgef' Lss=32e-9
+ Ldd=32e-9  Kgate='Kox' Tox='Hox' Csub='Cb' Vfbn='Vfn' Dout=0
+ Sout=0  Pitch=20e-9  n1=19  n2=0  tubes=nottube

XCNT9 OUT B N4 N4 NCNFET Lch=Lg  Lgeff='Lgef' Lss=32e-9
+ Ldd=32e-9  Kgate='Kox' Tox='Hox' Csub='Cb' Vfbn='Vfn' Dout=0
+ Sout=0  Pitch=20e-9  n1=10  n2=0  tubes=nottube

XCNT10 N4 A Source Sub NCNFET Lch=Lg  Lgeff='Lgef' Lss=32e-9
+ Ldd=32e-9  Kgate='Kox' Tox='Hox' Csub='Cb' Vfbn='Vfn' Dout=0
+ Sout=0  Pitch=20e-9  n1=10  n2=0  tubes=nottube

C1 OUT Gnd 0.5f

.ENDS TNAND
```

A.4.2 Two Input Ternary NOR Gate

In the code below, $OUT = \overline{A + B}$

```
.SUBCKT TNOR A B OUT Source Sub Vdd

XCNT1 N1 A Vdd Vdd PCNFET Lch=Lg  Lgeff='Lgef' Lss=32e-9
+ Ldd=32e-9  Kgate='Kox' Tox='Hox' Csub='Cb' Vfbn='Vfn' Dout=0
+ Sout=0  Pitch=20e-9  n1=19  n2=0  tubes=nottube

XCNT2 N2 B N1 N1 PCNFET Lch=Lg  Lgeff='Lgef' Lss=32e-9
+ Ldd=32e-9  Kgate='Kox' Tox='Hox' Csub='Cb' Vfbn='Vfn' Dout=0
+ Sout=0  Pitch=20e-9  n1=19  n2=0  tubes=nottube

XCNT3 N3 A Vdd Vdd PCNFET Lch=Lg  Lgeff='Lgef' Lss=32e-9
+ Ldd=32e-9  Kgate='Kox' Tox='Hox' Csub='Cb' Vfbn='Vfn' Dout=0
+ Sout=0  Pitch=20e-9  n1=10  n2=0  tubes=nottube
```

```
XCNT4 OUT B N3 N3 PCNFET Lch=Lg  Lgeff='Lgef' Lss=32e-9
+ Ldd=32e-9  Kgate='Kox' Tox='Hox' Csub='Cb' Vfbn='Vfn' Dout=0
+ Sout=0  Pitch=20e-9  n1=10  n2=0  tubes=nottube

XCNT5 OUT OUT N2 N2 PCNFET Lch=Lg  Lgeff='Lgef' Lss=32e-9
+ Ldd=32e-9  Kgate='Kox' Tox='Hox' Csub='Cb' Vfbn='Vfn' Dout=0
+ Sout=0  Pitch=20e-9  n1=13  n2=0  tubes=nottube

XCNT6 OUT OUT N4 N4 NCNFET Lch=Lg  Lgeff='Lgef' Lss=32e-9
+ Ldd=32e-9  Kgate='Kox' Tox='Hox' Csub='Cb' Vfbn='Vfn' Dout=0
+ Sout=0  Pitch=20e-9  n1=13  n2=0  tubes=nottube

XCNT7 N4 B Source Sub NCNFET Lch=Lg  Lgeff='Lgef' Lss=32e-9
+ Ldd=32e-9  Kgate='Kox' Tox='Hox' Csub='Cb' Vfbn='Vfn' Dout=0
+ Sout=0  Pitch=20e-9  n1=19  n2=0  tubes=nottube

XCNT8 N4 A Source Sub NCNFET Lch=Lg  Lgeff='Lgef' Lss=32e-9
+ Ldd=32e-9  Kgate='Kox' Tox='Hox' Csub='Cb' Vfbn='Vfn' Dout=0
+ Sout=0  Pitch=20e-9  n1=19  n2=0  tubes=nottube

XCNT9 OUT B Source Sub NCNFET Lch=Lg  Lgeff='Lgef' Lss=32e-9
+ Ldd=32e-9  Kgate='Kox' Tox='Hox' Csub='Cb' Vfbn='Vfn' Dout=0
+ Sout=0  Pitch=20e-9  n1=10  n2=0  tubes=nottube

XCNT10 OUT A Source Sub NCNFET Lch=Lg  Lgeff='Lgef' Lss=32e-9
+ Ldd=32e-9  Kgate='Kox' Tox='Hox' Csub='Cb' Vfbn='Vfn' Dout=0
+ Sout=0  Pitch=20e-9  n1=10  n2=0  tubes=nottube

C1 OUT Gnd 0.5f

.ENDS TNOR
```

A.4.3   Two Input Ternary AND Gate

In the code below, $OUT = A \cdot B$.

```
.SUBCKT TAND A B OUT Source Sub Vdd

XTNAND A B Y Source Sub Vdd TNAND
XINV Y OUT Source Sub Vdd STDINV
```

```
C1 OUT Gnd 0.5f

.ENDS TAND
```

A.4.4   Two Input Ternary OR Gate

In the code below, $OUT = A + B$

```
.SUBCKT TOR A B OUT Source Sub Vdd

XTNOR A B Y Source Sub Vdd TNOR
XINV Y OUT Source Sub Vdd STDINV

C1 OUT Gnd 0.5f

.ENDS TOR
```

A.5   HSPICE Code for Ternary Multiplexer

A.5.1   Ternary Transmission Gate

```
.SUBCKT TXGATE D S S_BAR OUT Source Vdd

XCNT1 OUT S D Vdd PCNFET Lch=Lg  Lgeff='Lgef' Lss=32e-9
+ Ldd=32e-9  Kgate='Kox' Tox='Hox' Csub='Cb' Vfbn='Vfn' Dout=0
+ Sout=0  Pitch=20e-9  n1=10  n2=0  tubes=nottube

XCNT2 OUT S_BAR D Gnd NCNFET Lch=Lg  Lgeff='Lgef' Lss=32e-9
+ Ldd=32e-9  Kgate='Kox' Tox='Hox' Csub='Cb' Vfbn='Vfn' Dout=0
+ Sout=0  Pitch=20e-9  n1=19  n2=0  tubes=nottube

.ENDS TXGATE
```

A.5.2 3 × 1 Ternary MUX

Here S is the select signal, $D0$, $D1$ and $D2$ are the data lines. All these three valued signals are specified as given below.

```
VD2 D2 Gnd PWL(0 0 '8*T' 0 '8*T+0.01p' 0.45 '8.5*T' 0.45
+ '8.5*T+0.01p' 0.9 '10*T' 0.9)


VD1 D1 Gnd PWL(0 0 '1*T' 0 '1*T+0.01p' 0.45 '2*T' 0.45
+ '2*T+0.01p' 0.9 '3*T' 0.9 '3*T+0.01p' 0.0 '6*T' 0.0)


VD0 D0 Gnd PWL(0 0 'T' 0 'T+0.01p' 0.45 '4*T' 0.45 '4*T+0.01p'
+ 0.9 '5*T' 0.9 '5*T+0.01p' 0.45 '6*T' 0.45 '6*T+0.01p' 0 '8*T'
+ 0 '8*T+0.01p' 0.9 '10*T' 0.9)


VS S Gnd PWL(0 0 '3*T' 0 '3*T+0.01p' 0.45 '7*T' 0.45
+ '7*T+0.01p' 0.9 '10*T' 0.9)
```

The code for the multiplexer itself is given next.

```
.SUBCKT TMUX S D0 D1 D2 OUT Source Sub Vdd

XPOSINV0 S   SP  Source Sub Vdd POSINV
XNEGINV1 S   SN  Source Sub Vdd NEGINV
XINV2        SP SPB Source Sub Vdd INV

XINV3    S S1 Source Sub Vdd Dec_A1
XINV4    S1 S1B Source Sub Vdd  INV

XTXGATE0 D0 S SN OUT Source Vdd TXGATE

XTXGATE1 D1 S1B S1 OUT Source Vdd TXGATE

XTXGATE2 D2 SP SPB OUT Source Vdd TXGATE

C1 OUT Gnd 0.5f

.ENDS TMUX
```

A.5.3 HSPICE Codes for Ternary 2 × 1 MUX

- Here S is the two valued signal which takes values of '0' or '1'. The data lines are three-valued signals as given below.

```
VD1 D1 Gnd PWL(0 0 '1*T' 0 '1*T+0.01p' 0.45 '2*T' 0.45
+ '2*T+0.01p' 0.9 '3*T' 0.9 '3*T+0.01p' 0.0 '6*T' 0.0)

VD0 D0 Gnd PWL(0 0 'T' 0 'T+0.01p' 0.45 '4*T' 0.45 '4*T+0.01p'
+ 0.9 '5*T' 0.9 '5*T+0.01p' 0.45 '6*T' 0.45 '6*T+0.01p' 0 '8*T'
+ 0 '8*T+0.01p' 0.9 '10*T' 0.9)

VS S Gnd PWL(0 0 '3*T' 0 '3*T+0.01p' 0.45 '7*T' 0.45
+ '7*T+0.01p' 0 '10*T' 0)
```

The code for the Ternary 2 × 1 multiplexer is given next.

```
.SUBCKT TMUX_2x1_01 S D0 D1 OUT Source Sub Vdd

XNEGINV1 S   SN  Source Sub Vdd NEGINV
XINV2    SN SNB Source Sub Vdd INV

XTXGATE0 D0 S SN OUT Source Vdd TXGATE

XTXGATE1 D1 SN SNB OUT Source Vdd TXGATE

C1 OUT Gnd 0.5f

.ENDS TMUX_2x1_01
```

We now present the code for the second type of 2 × 1 multiplexer. That is, we assume that the select signal takes values of 1 or 2. The data lines are three valued signals.

```
VD1 D1 Gnd PWL(0 0 '1*T' 0 '1*T+0.01p' 0.45 '2*T' 0.45
+ '2*T+0.01p' 0.9 '3*T' 0.9 '3*T+0.01p' 0.0 '6*T' 0.0)

VD2 D2 Gnd PWL(0 0 'T' 0 'T+0.01p' 0.45 '4*T' 0.45 '4*T+0.01p'
```

```
+ 0.9 '5*T' 0.9 '5*T+0.01p' 0.45 '6*T' 0.45 '6*T+0.01p' 0 '8*T'
+ 0 '8*T+0.01p' 0.9 '10*T' 0.9)

VS S Gnd PWL(0 0.45 '3*T' 0.45 '3*T+0.01p' 0.9 '7*T' 0.9
+ '7*T+0.01p' 0.45 '10*T' 0.45)
```

The code for this ternary 2 × 1 multiplexer is given next.

```
.SUBCKT TMUX_2x1_12 S D1 D2 OUT Source Sub Vdd

XPOSINV1 S  SP  Source Sub Vdd POSINV
XINV2     SP SPB Source Sub Vdd INV

XTXGATE0 D0 SPB SP OUT Source Vdd TXGATE

XTXGATE1 D1 SP S OUT Source Vdd TXGATE

C1 OUT Gnd 0.5f

.ENDS TMUX_2x1_12
```

We now consider the third possibility: S is the two valued signal which takes values of '0' or '2'. The data lines are three valued signals as given below.

```
VD2 D2 Gnd PWL(0 0 '1*T' 0 '1*T+0.01p' 0.45 '2*T' 0.45
+ '2*T+0.01p' 0.9 '3*T' 0.9 '3*T+0.01p' 0.0 '6*T' 0.0)

VD0 D0 Gnd PWL(0 0 'T' 0 'T+0.01p' 0.45 '4*T' 0.45 '4*T+0.01p'
+ 0.9 '5*T' 0.9 '5*T+0.01p' 0.45 '6*T' 0.45 '6*T+0.01p' 0 '8*T'
+ 0 '8*T+0.01p' 0.9 '10*T' 0.9)

VS S Gnd PWL(0 0 '3*T' 0 '3*T+0.01p' 0.9 '7*T' 0.9
+ '7*T+0.01p' 0 '10*T' 0)
```

The code for the ternary 2 × 1 multiplexer in this case follows.

```
.SUBCKT TMUX_2x1_02 S D0 D2 OUT Source Sub Vdd

XNEGINV1 S  SN  Source Sub Vdd NEGINV
XINV2     SN SNB Source Sub Vdd INV
```

```
XTXGATE0 D0 S SN OUT Source Vdd TXGATE

XTXGATE1 D1 SN S OUT Source Vdd TXGATE

C1 OUT Gnd 0.5f

.ENDS TMUX_2x1_02
```

A.6   HSPICE Code for Ternary Half Adder

The inputs for a ternary half-adder are specified first.

```
VA A Gnd PWL(0.0 0.0 '3*T' 0 '3*T+0.01p' 0.45 '6*T' 0.45
+ '6*T+0.01p' 0.9 '9*T' 0.9 '9*T+0.01p') **0.0 '9*T' 0.0
+ '9*T+0.01p' 0.9 '10*T' 0.9)

VB B Gnd PWL(0.0 0.9 '3*T' 0.9 '3*T+0.01p' 0.45 '4*T' 0.45
+ '4*T+0.01p' 0.0 '6*T' 0.0 '6*T+0.01p' 0.9 '7*T' 0.9
+ '7*T+0.01p' 0.0 '8*T' 0.0 '8*T+0.01p' 0.9 '10*T' 0.9)
```

In addition to the given code for ternary half adder, we need to copy all the required sub-circuits from the Sect. A.3 and the portions from Sect. A.1. The specific code for the half-adder follows.

```
.SUBCKT THA A B SUM CARRY Source Sub Vdd Vmm

XCKT1 A Y1 Source Sub Vdd cycle_A1
XCKT2 A Y2 Source Sub Vdd cycle_A2

XCKT3 A Y3 Source Sub Vdd CKT_001
XCKT4 A Y4 Source Sub Vdd CKT_011

XTMUX0 B  A   Y1  Y2  SUM Source Sub Vdd TMUX
XTMUX1 B  Gnd Y3  Y4  CARRY Source Sub Vdd TMUX

C1 SUM  Gnd 0.5f
C2 CARRY Gnd 0.5f

.ENDS THA
```

A.7 HSPICE Code for Ternary Full Adder

The inputs for a ternary full-adder are specified first.

```
VA A Gnd PWL(0.0 0.0 '3*T' 0 '3*T+0.01p' 0.45 '6*T' 0.45
+ '6*T+0.01p' 0.9 '9*T' 0.9 '9*T+0.01p') **0.0 '9*T' 0.0
+ '9*T+0.01p' 0.9 '10*T' 0.9)


VB B Gnd PWL(0.0 0.9 '3*T' 0.9 '3*T+0.01p' 0.45 '4*T' 0.45
+  '4*T+0.01p' 0.0 '6*T' 0.0 '6*T+0.01p' 0.9 '7*T' 0.9
+ '7*T+0.01p' 0.0 '8*T' 0.0 '8*T+0.01p' 0.9 '10*T' 0.9)


VC C Gnd PWL(0.0 0.9 '1*T' 0.9 '1*T+0.01p' 0.45 '2*T' 0.45
+  '2*T+0.01p' 0.0 '4*T' 0.0 '4*T+0.01p' 0.45 '5*T' 0.45
+ '5*T+0.01p' 0.9 '7*T' 0.9 '7*T+0.01p' 0.0 '8*T'
+ 0.0 '8*T+0.01p' 0.45  '9*T' 0.45 '9*T+0.01p' 0.9 '10*T' 0.9)
```

In addition to the given code for ternary full adder, we need to copy all the required sub circuits from the Sect. A.3 and the portions from Sect. A.1. The adder-specific code follows.

```
.SUBCKT TFA A B C SUM CARRY Source Sub Vdd Vmm

XCKT1 A Y1 Source Sub Vdd cycle_A1
XCKT2 A Y2 Source Sub Vdd cycle_A2
XCKT3 A Y3 Source Sub Vdd CKT_001
XCKT4 A Y4 Source Sub Vdd CKT_011
XCKT5 A Y5 Source Sub Vdd CKT_112


XTMUX0 B  A    Y1  Y2  PS0 Source Sub Vdd TMUX
XTMUX1 B  Gnd  Y3  Y4  PC0 Source Sub Vdd TMUX


XTMUX2 B  Y1  Y2  A    PS1 Source Sub Vdd TMUX
XTMUX3 B  Y4  Y3  Vmm  PC1 Source Sub Vdd TMUX


XTMUX4 B  Y2  A    Y1  PS2 Source Sub Vdd TMUX
XTMUX5 B  Y3  Vmm  Y4  PC2 Source Sub Vdd TMUX


XTMUX6 C PS0 PS1 PS2 SUM   Source Sub Vdd TMUX
XTMUX7 C PC0 PC1 PC2 CARRY Source Sub Vdd TMUX
```

```
C1  SUM   Gnd 0.5f
C2  CARRY Gnd 0.5f


.ENDS TFA
```

A.8   HSPICE Code for Two-Input Ternary Multiplier

The inputs for a ternary multiplier are specified as below.

```
VA A Gnd PWL(0.0 0.0 '3*T' 0 '3*T+0.01p' 0.45 '6*T' 0.45
+ '6*T+0.01p' 0.9 '9*T' 0.9 '9*T+0.01p') **0.0 '9*T' 0.0
+ '9*T+0.01p' 0.9 '10*T' 0.9)

VB B Gnd PWL(0.0 0.9 '3*T' 0.9 '3*T+0.01p' 0.45 '4*T' 0.45
+ '4*T+0.01p' 0.0 '6*T' 0.0 '6*T+0.01p' 0.9 '7*T' 0.9
+ '7*T+0.01p' 0.0 '8*T' 0.0 '8*T+0.01p' 0.9 '10*T' 0.9)
```

In addition to the given code for ternary multiplier, we need to copy all the required sub circuits from the Sect. A.3 and the portions from Sect. A.1 to run the transient analysis for the circuit. The remaining code is as follows.

```
.SUBCKT TMUL A B PROD CARRY Source Sub Vdd Vmm


XCKT1 A Y1 Source Sub Vdd CKT_021
XCKT2 A Y2 Source Sub Vdd CKT_001


XTMUX0 B  Gnd A   Y1  PROD  Source Sub Vdd TMUX
XTMUX1 B  Gnd Gnd Y2  CARRY Source Sub Vdd TMUX


C1 PROD  Gnd 0.5f
C2 CARRY Gnd 0.5f


.ENDS TMUL
```

A.9 Summary

In this appendix, we have presented HSPICE code for various circuits. It is worth noting that the code for an arithmetic circuit that performs addition or multiplication can be composed from the code for the various primitives.

Appendix B
Python Code for Automatic Synthesis

In this appendix, we present Python code developed for automatic synthesis of ternary logic circuits. Various functions for which the synthesis procedure can be applied are defined here. Sample output is also shown (along with user inputs).

B.1 The Program

```
### The algorithm can be applied to these
## functions upon selection###

def tand(list):    #### Ternary AND ####
    return min(list)

def tor(list): #### Ternary OR ####
    return max(list)

def nand(list): #### Ternary NAND ####
    return 2-min(list)

def nor(list): #### Ternary NOR ####
    return 2-max(list)

def product(A): #### Ternary Multiplication ####
    p = 1
    for i in list(A):
        p *= i
    return p

def MAC(A): #### Ternary Multiply and Accumulate ####
```

© Springer Nature Switzerland AG 2020
K. Sridharan et al., *Low-Complexity Arithmetic Circuit Design in Carbon Nanotube Field Effect Transistor Technology*, Carbon Nanostructures,
https://doi.org/10.1007/978-3-030-50699-5

```
        D=[]
    for i in range(len(A)):
        D=A[0]*A[1] + A[2]
    return D
########## Decimal to ternary conversion ###########
def ternary_one(n):
        num=[]
        x=n;
        while (x > 2):
            a = x/3
            b = x%3
            x=a
            num.insert(0,b)
        else:
            num.insert(0,x)

        return num
########## Generate the 'k' digit ternary numbers ##########
def ternary(k):
        m=pow(3,k)
        n=range(0,m)
        Y = []
        T=[]
        for i in range (m):
          num=[]
          x=n[i];
          while (x > 2):
            a = x/3
            b = x%3
            x=a
            num.insert(0,b)
          else:
            num.insert(0,x)
          Y.append(num)
        F=len(Y)
        G=len(Y[F-1])
        for k in range(0,F):
            R=[]
            H=len(Y[k])
            l=G-H
            if(H==G):
                R.extend(Y[k])
            else:
                R=[0]*l
                R.extend(Y[k])
```

```
            T.append(R)
        return T


############## Choose a particular function  #########

def ter_fun(n,r,fun):
        A=ternary(n)
        print(A)
        m=pow(3,n)
#       P=[]
        Q=[]
#       R=[]
        S=[]
        T=[]
        for i in range(m):
            if(fun==0):
                S.append(sum(A[i]))
            elif(fun==1):
                S.append(product(A[i]))
            elif(fun==2):
                S.append(nand(A[i]))
            elif(fun==3):
                S.append(nor(A[i]))
            elif(fun==4):
                S.append(tand(A[i]))
            elif(fun==5):
                S.append(tor(A[i]))
            elif(fun==6):
                S.append(MAC(A[i]))
            else:
                    S.append(counter(A[i]))

            Q.append(ternary_one(S[i]))
        print ('Output is :',S)
        F=len(Q)
        G=len(Q[F-1])
        for k in range(0,F):
            R=[]
            H=len(Q[k])

            l=G-H
            if(H==G):
                    R.extend(Q[k])
            else:
                    R=[0]*l
```

```
                R.extend(Q[k])
          T.append(R)
     return T


############ Generate the outputs in order ################
def ter_outputs(Q):
    T=[]
    a=len(Q) # number of outputs
    b=len(Q[0])  # number of bits
    for j in range(b):
        Y=[]
        for i in range(a):
          Y.append(Q[i][j])

          T.append(Y)
    return T


#########Count the number of 0's, 1's and 2's ###########
def count(X):
    A=X.count(0)
    B=X.count(1)
    C=X.count(2)
    Y=[A,B,C]
    return Y


######### Generate the decomposition output ##########
def Fx(i,j,X):
    Y=[]
    for k in range(i,j+1):
        Y.append(X[k])
    return Y


####Main algorithm######
def algo_decom(n,X):
    y=len(X)
    z=len(X[y-1])
    for i in range(y):
        A=X[i]
#### Check whether the output is a constant ####
        if(sum(A[:])==0):
            print("Complete zero function")
            done=1
            k=0
            e=0
            Qq=0
```

```
                dne=1
          elif(A[:]==[1]*z):
               print("complete one function")
               done=1
               e=0
               k=0
               Qq=0
               dne=1
          elif(A[:]==[2]*z):
               print("complete two function")
               done=1
               e=0
               k=0
               Qq=0
               dne=1
          else:
               print("Case 1 failed :: Decomposition required")
               Qq=1
               dne=0
          if(dne==1):
             Y=[]
             P=A[0:(z/3)]
             Q=A[(z/3):2*(z/3)]
             R=A[2*(z/3):z]
             Y.append(P)
             Y.append(Q)
             Y.append(R)
             done=1
       if(Qq==1):
               print("Now checking which output to consider")
               for i in range(y):
                    D=[]
                    D.append((X[i]).count(0))
                    e=D.index(max(D))
                    B=X[e]
                    C0=X[e].count(0)
                    C1=X[e].count(1)
                    C2=X[e].count(2)
                    k=1
                    l=0
                    dec=0
                    y=len(X[e])
#### Checking whether one third of output is constant ####
                    if(((C0>=y/3) or (C1>=y/3) or (C2>=y/3))):
                         for o in range(n+1):
```

```
                              if(dec<n):
                                  P=[]
                                  Q=[]
                                  R=[]
                                  i=0
                                  j=y/pow(3,k)
                                  jj=j
                                  mm=pow(3,l)
                                  for m in range(0,mm):
                                   ## P : decomposed along C ##
                                      P.extend(Fx(i,(j-1),B))

                                   ## Q : decomposed along A ##
                                      Q.extend(Fx(i,(j-1),B))

                                   ## R : decomposed along B ##
                                      R.extend(Fx(i,j-1,B))
                                      i=j
                                      j=i+jj
                  ## Checking whether any layer (P, Q or R) is constant ###
                                  if(sum(P[:])==0  or TAND(P,1)==1 \
        or TAND(P,2)==2 or sum(Q[:])==0 \
         or TAND(Q,1)==1  or TAND(Q,2)==2 \
          or sum(R[:])==0 or TAND(R,1)==1 \
          or TAND(R,2)==2) :
                                      if(sum(P[:])==0  or TAND(P,1)==1 \
        or TAND(P,2)==2):
                                          print("Decompose along I_k",k)
                                      if(sum(Q[:])==0  or TAND(Q,1)==1 \
        or TAND(Q,2)==2):
                                          print("Decompose along I_k",k)
                                      if(sum(R[:])==0  or TAND(R,1)==1 \
        or TAND(R,2)==2):
                                          print("Decompose along I_k",k)
                                          Y=[P,Q,R]
                                          done=1
                                          k=k
                                          dec=dec+n
                                          Qq=0
                                  else:
                                      k=k+1
                                      l=l+1
                                      dec+=1
                          else:
                              sym=1
```

```
                            Qq=1
                            dec=0
                            k=1
                            l=0
                    else:
                        sym=1
                    if(sym==1):
                        print('Now, have to check for symmetry')
                            sym=1
                            dec=0
                            k=1
                            l=0
                            P=[]
                            Q=[]
                            R=[]
                            i=0
                            j=y/pow(3,k)
                            jj=j
                            mm=pow(3,l)
                            P.extend(Fx(i,(j-1),B))
                            i=j
                            j=i+jj
                            Q.extend(Fx(i,(j-1),B))
                            i=j
                            j=i+jj
                            R.extend(Fx(i,(j-1),B))
                            i=j
                            j=i+jj
                            Y=[P,Q,R]
                            for o in range(n):
                                if(dec<(n-1)):
                                print('Checking for symmetry  ')
                                    if(sym==1):
                                        T=[]
                                        L=[]
            #### First check the count of P, Q and R #####
                                    if(count(P) == count(Q) \
== count(R)):
    ### Checking the similarity between P, Q and R#####
                                        if(symmetry(P,Q,R) == True ):
                                            print("Symmetry matched")
                                    print("Decompose along I_k ",k)
                                            dec=dec+n
                                            sym=0
                                            done=1
```

```
                                    Qq=0
                                    k=k
                                    Y=[P,Q,R]
                                else:
                                    dec=dec+1
                                    T.append([k])
                                    L.append([l])
                           else:
                               print("No symmetry")
                               k=k+1
                               l=l+1
                               dec+=1
                               i=0
                               j=y/pow(3,k)
                               jj=j
                               mm=pow(3,l)
                           for m in range(0,mm):
                         P.append(Fx(i,(j-1),B))
                                   i=j
                                   j=i+jj
                          Q.append(Fx(i,(j-1),B))
                                   i=j
                                   j=i+jj
                          R.append(Fx(i,(j-1),B))
                                   i=j
                                   j=i+jj
                               sym=1
                 else:
                   print('Symmetry has failed')
                   done=1
                   Qq=0
                   P=[]
                   Q=[]
                   R=[]
                   k=1
                   l=0
                   i=0
                   j=y/pow(3,k)
                   jj=j
                   mm=pow(3,l)
                   for m in range(0,mm):
                       P.extend(Fx(i,(j-1),B))
                       i=j
                       j=i+jj
                       Q.extend(Fx(i,(j-1),B))
```

```
                                        i=j
                                        j=i+jj
                                        R.extend(Fx(i,(j-1),B))
                                        i=j
                                        j=i+jj
                                 Y=[P,Q,R]
                                 print("Along Ik direction",k)
                                 print("Decompose along I_k ",k)
                                 done=1
                                 Qq=0
                                 k=k
        if(done==1):
                G=[]
                y=len(X)
                h=len(X)
                for i in range(h):
                        if(i==e):
                            G.extend(Y)
                        else:
                            B=X[i]
                            P=[]
                            Q=[]
                            R=[]
                            l=0
                            i=0
                            y=len(B)
                            j=y/pow(3,k)
                            jj=j
                            mm=pow(3,l)
                            for m in range(0,mm):
                                    P.extend(Fx(i,(j-1),B))
                                        i=j
                                        j=i+jj
                                    Q.extend(Fx(i,(j-1),B))
                                        i=j
                                        j=i+jj
                                    R.extend(Fx(i,(j-1),B))
                                        i=j
                                        j=i+jj
                            Y=[P,Q,R]
                            G.extend(Y)
                return G


n= input('enter no. of inputs : ')
```

```
fun=input('enter function  :: \n [0=sum],[1=prod], \
\n [2=nand],[3=nor],\
\n [4=and],[5=or], \n \
[6: MAC] ::')
r= input('enter no. of outputs :')
Y=ter_fun(n,r,fun)
Z=ter_outputs(Y)

W=algo_decom(n,Z)

print('Decomposed layers::')
print '\n'.join(map(str, W))
print("Decomposition is Done")
```

B.2   Sample Output

```
enter n : 3
enter function ::
 [0=sum],[1=prod],
 [2=nand],[3=nor],
 [4=and],[5=or],
 [6: MAC] ::0
enter r:2
[[0, 0, 0], [0, 0, 1], [0, 0, 2], [0, 1, 0], [0, 1, 1],
 [0, 1, 2], [0, 2, 0], [0, 2, 1], [0, 2, 2], [1, 0, 0],
 [1, 0, 1], [1, 0, 2], [1, 1, 0], [1, 1, 1], [1, 1, 2],
 [1, 2, 0], [1, 2, 1], [1, 2, 2], [2, 0, 0], [2, 0, 1],
 [2, 0, 2], [2, 1, 0], [2, 1, 1], [2, 1, 2], [2, 2, 0],
 [2, 2, 1], [2, 2, 2]]
('Output is :', [0, 1, 2, 1, 2, 3, 2, 3, 4, 1, 2, 3, 2,
 3, 4, 3, 4, 5, 2, 3, 4, 3, 4, 5, 4, 5, 6])
Case 1 failed :: Decomposition required
Case 1 failed :: Decomposition required
Now checking which output to consider
Now, have to check for symmetry
Checking for symmetry
No symmetry
Checking for symmetry
No symmetry
Symmetry has failed
```

```
('Along Ik direction', 1)
('Decompose along I_k ', 1)
Now, have to check for symmetry
Checking for symmetry
No symmetry
Checking for symmetry
No symmetry
Symmetry has failed
('Along Ik direction', 1)
('Decompose along I_k ', 1)
Decomposed layers::
[0, 0, 0, 0, 0, 1, 0, 1, 1]
[0, 0, 1, 0, 1, 1, 1, 1, 1]
[0, 1, 1, 1, 1, 1, 1, 1, 2]
[0, 1, 2, 1, 2, 0, 2, 0, 1]
[1, 2, 0, 2, 0, 1, 0, 1, 2]
[2, 0, 1, 0, 1, 2, 1, 2, 0]
Decomposition is Done
```

B.3 Summary

In this appendix, we have presented Python code for automatic synthesis. We have focussed on three variable functions.

Index

© Springer Nature Switzerland AG 2020
K. Sridharan et al., *Low-Complexity Arithmetic Circuit Design in Carbon Nanotube Field Effect Transistor Technology*, Carbon Nanostructures,
https://doi.org/10.1007/978-3-030-50699-5

Printed in the United States
by Baker & Taylor Publisher Services